EXAMPLES OF MATLAB 5, CONTROL SYSTEM AND RPI-FUNCTION COMMANDS

```
ph = angle(x)                            % phase angle of a complex array
axis equal                               % for uniform scaling on both axes
axis normal                              % for default scaling on both axes
axis([-20 5 -10 10])                     % specify plotting region
bode(G)                                  % draw Bode plot for LTI object G
[mag,ph] = bode(G,w)                     % compute mag and phase, no plot
bw = bwcalc(mag_db,ww,lfg_db)            % calculate bandwidth
Gm = canon(G,'modal')                    % real modal form of LTI object G
clear                                    % remove all variables from workspace
clf                                      % clear current graphics window
denG = conv([tauP1 1],[tauP2 1])         % multiply two polynomials
ycmplx = cpole2t(polG,resG,t)            % response due to complex poles
Co = ctrb(G)                 % compute controllability matrix of SS object
damp(denG)                               % compute damping ratio
damp(G)                  % compute damping ratio of poles of LTI object G
plant_lfg = dcgain(G)                    % DC gain of LTI object
quo = deconv(num,den)          % divide polynomial num by polynomial den
disp('Problem 2.1')                      % display text
eig(A)                                   % eigenvalues of A
T = feedback(G,H)                        % cl-loop system of two LTI objects
T = feedback(G,1)                        % cl-loop system, unity feedback
figure(4)                                % make figure window 4 current
k = find((ph <= -100) & (ph >= -120))    % indices of selected elements
for k = 1:10, ..., end                   % for loop
format                                   % default fixed point format
format short e                           % floating point format
grid                                     % plot grid lines
hold on, ..., hold off                   % freeze, then release plot
if k > 3, ..., end                       % execute conditional statement
impulse(G)                     % plot impulse response of LTI object G
y = impulse(G,t)         % impulse response, with specified time array t
KI = input('enter integral gain KI => ')       % user specifies KI
wc = interp1(mag,w,mag_des)              % interpolate to find wc
keyboard                                 % invoke keyboard
length(t)                                % length of vector t
mag_db = 20*log10(mag)                   % convert magnitude to decibels
w = logspace(-1,2)       % log-spaced values from 1e-1 to 1e2, 50 points
w = logspace(-1,2,100)                   % 100 points
lsim(G,u,t)            % plot LTI object response to general input u, no IC
[y,t,x] = lsim(G,u,t,x0)     % SS object response to general input with IC
margin(G)                      % find gain and phase margins and plot
[km,pm,wg,wp] = margin(G)                % find gain and phase margins
ngrid                                    % add Nichols grid lines
```

```
nichols(G,w)          % Nichols plot of LTI object G with freq specified
y = norm(x)                                  % norm of vector or matrix
x = num2str(1.23)                          % convert number to string
nyquist(G)       % Nyquist plot of LTI object G where MATLAB selects freq
Ob = obsv(G)                            % compute observability matrix
G = G1+G2                  % parallel conn, using overloaded operator +
pause                                        % wait for user response
F = place(A,B,p)                % find control gains for pole placement
plot(t,y_ref,t,y_dist,'--')                % plot 2 curves, second dashed
pzmap(G)                          % plot zeros & poles of LTI object G
[p,z] = pzmap(G)                     % output poles to p and zeros to z
rank(A)                                             % rank of a matrix
[resG,polG,otherG] = residue(numG,denG)      % residues & poles of G(s)
return                                    % return to invoking function
[kk,pCL] = rlocfind(G)               % calibrate RL plot with gain values
rlocus(G)                           % plot root locus for LTI object G
p = roots(denG)                                        % poles of G(s)
yreal = rpole2t(polG,resG,t)              % response due to a real pole
semilogx(w,20*log10(mag))              % semi-logarithmic magnitude plot
G = G1*G2                    % series conn, using overloaded operator *
sgrid(0.8,[ ])                      % draw lines for damping ratio = 0.8
sgrid([ ],[-1 1])                  % draw unit circle centered at origin
y = small20(x)                   % y = x with small elements set to zero
G = ss(A,B,C,D)                                    % create SS object
G = ss(H)                             % convert LTI object H to SS object
[A,B,C,D] = ssdata(G)                          % retrieve SS model data
step(G)                           % generate step response of LTI object G
step(G,t)          % generate step response, with specified time array
subplot(2,1,1)             % partition figure window and set to first plot
text(3,5,'Kp = 10')              % put text string at coordinate (3,5)
title('Step response')                                % title for plot
G = tf(numG,denG)                                  % create TF object
G = tf(H)                             % convert LTI object H to TF object
[numG,denG] = tfdata(G,'v')                    % retrieve TF model data
[Mo,tp,tr,ts,ess] = tstats(t,y,yss)    % find time specs for step response
z = tzero(G)                        % find system zeros of LTI object G
x = vec2str([1.2 3.4])                  % convert row vector to string
Kv = vgain(G)                          % find velocity error constant
while(k > 0), ..., end                               % repeat statement
xlabel('Time (secs)'), ylabel('Amplitude')   % label for x-axis and y-axis
G = zpk(zz,pp,kk)                                  % create ZPK object
G = zpk(H)                          % convert LTI object H to ZPK object
[zz,pp,kk] = zpkdata(G,'v')                   % retrieve ZPK model data
numG = [3 2], denG = [2 4 5 1]                     % defining variables
t = [0:0.02:10]'                      % time points as column vector
```

BookWare Companion Series™

Feedback Control Problems

Using MATLAB® and the Control System Toolbox

Dean K. Frederick
Unified Technologies, Inc.

Joe H. Chow
Rensselaer Polytechnic Institute

Pacific Grove • Albany • Belmont • Boston • Cincinnati • Johannesburg • London • Madrid • Melbourne
Mexico City • New York • Scottsdale • Singapore • Tokyo • Toronto

Publisher: *Bill Stenquist*
Marketing Team: *Christina DeVeto and Nathan Wilbur*
Editorial Assistant: *Shelley Gesicki*
Production Editor: *Mary Vezilich*
Production Service: *Greg Hubit Bookworks*
Cover Design: *Denise Davidson*
Print Buyer: *Jessica Reed*
Typesetting: *The Beacon Group, Inc.*
Printing and Binding: *Webcom Ltd.*

For more information, contact:
BROOKS/COLE
511 Forest Lodge Road
Pacific Grove, CA 93950 USA
www.brookscole.com

Printed in Canada

10 9 8 7 6 5 4 3 2

Library of Congress Cataloging-in-Publication Data

Frederick, Dean K.
Feedback control problems : using MATLAB and the Control System Toolbox / Dean K. Frederick, Joe H. Chow.
p. cm. — (BookWare companion series)
Includes bibliographical references.
ISBN 0-534-37172-8
1. Feedback control systems. 2. MATLAB. 3. Control system toolbox (Computer file) I. Chow, J. H. (Joe H.).
II. Title. III. Series: BookWare companion series (Pacific Grove, Calif.)
TJ216.F73 2000
629.8′3—dc21 99-41227
CIP

Books in the BookWare Companion Series™

Djaferis	*Automatic Control: The Power of Feedback Using MATLAB®*
Frederick/Chow	*Feedback Control Problems Using MATLAB® and the Control System Toolbox*
Ingle/Proakis	*Digital Signal Processing Using MATLAB®*
Pfeiffer	*Basic Probability Topics Using MATLAB®*
Proakis/Salehi	*Contemporary Communications Systems Using MATLAB®*
Rashid	*Electronics Circuit Design Using ELECTRONICS WORKBENCH*
Stonick/Bradley	*Labs for Signals and Systems Using MATLAB®*
Strum/Kirk	*Contemporary Linear Systems Using MATLAB®*

About the Series

"The purpose of computing is insight, not numbers."
—R. W. Hamming, *Numerical Methods for Engineers and Scientists,* McGraw-Hill, Inc.

It is with this spirit in mind that we present the BookWare Companion Series.™

Increasingly, the latest technologies and modern methods are crammed into courses already dense with important theory. As a result, many instructors now ask, "Are we simply teaching students the latest technology, or are we teaching them to reason?" We believe that these two alternatives need not be mutually exclusive. In fact, this series was founded on the belief that computer solutions and theory can be mutually reinforcing. Properly applied, computing can illuminate theory and help students to think, analyze, and reason in meaningful ways. It can also help them to understand the relationships and connections between new information and existing knowledge and to cultivate problem-solving skills, intuition, and critical thinking. The BookWare Companion Series was developed in response to this mission.

Specifically, the series is designed for educators who want to integrate computer-based learning tools into their courses, and for students who want to go further than their textbook alone allows. The former will find in the series the means by which to use powerful software tools to support their course activities without having to customize the applications themselves. The latter will find relevant problems and examples quickly and easily available and will have electronic access to them. Important for both educators and students is the premise on which the series is based: that students learn best when they are actively involved in their own learning. The BookWare Companion Series will engage them, provide a taste of real-life issues, demonstrate clear techniques for solving real problems, and challenge them to understand and apply these techniques on their own.

To serve your needs better, we are continually looking for ways to improve the series. Toward that end, please join us at our BookWare Companion Resource Center website:

http://www.brookscole.com/engineering/ee/bookware.html

You can recommend ways to make the series even better, share your ideas about using technology in the classroom with your colleagues, suggest a specific problem or example for the next edition, or just let us know what's on your mind. We look forward to hearing from you, and we thank you for your continuing support.

Bill Stenquist	Publisher	bill.stenquist@brookscole.com
Shelley Gesicki	Editorial Assistant	shelley.gesicki@brookscole.com
Nathan Wilbur	Marketing Manager	nathan.wilbur@brookscole.com
Christina DeVeto	Marketing Assistant	christina.deveto@brookscole.com

To my grandchildren Miles, Charlotte, Danielle,
Taylor, Keltin, and Jack,
and to those yet to come
DKF

To Doris, Amy, and Tammy
JHC

Contents

1 Introduction

2 Single-block Models and Their Responses

5 Root-locus Plots

6 Frequency-response Analysis

CONTENTS

8 Proportional-Integral-Derivative Control

9 Frequency-response Design

10 State-space Design Methods

A Models of Practical Systems

B MATLAB Commands

Preface

The purpose of this book is to assist those who are studying the introductory aspects of control systems engineering by allowing them to use a digital computer to rapidly work a wide range of numerical problems so as to reinforce the learning process. The book is built around illustrative examples that demonstrate the steps involved in the analysis and design process. The examples are followed by a variety of problems that span the spectrum from follow-up what-if problems, to simple textbook-type reinforcement problems, to open-ended exploratory problems, and to realistic comprehensive problems.

To accomplish this objective, this book uses the power of MATLAB and its Control System Toolbox. It is anticipated that this book and the accompanying files that can be downloaded from the Brooks/Cole web site will be used as a supplement to one of the numerous textbooks that cover the introductory aspects of feedback control. It is not intended that this book be used as the sole reference for learning this material, as key results are summarized and illustrated but are not developed from basic principles.

The combination of this book and a suitable computer that runs MATLAB and the Control System Toolbox will provide the user with numerous opportunities for applying the techniques of linear system analysis that form the basis for the analysis and design of feedback control systems. Because a powerful computer environment is almost always available, the user is no longer restricted to solving first- and second-order problems that can be worked out by hand. Nor is the user subjected to the drudgery of performing the laborious calculations required to solve meaningful problems. Rather, the user can concentrate on interpreting the analysis and design results obtained with the computer, thereby enhancing the learning process.

The problems in this book are suitable for students taking a senior-level course on feedback control design of continuous-time systems. In addition to classical control design methods based on transfer-function models, the book also includes problems on state-space models and design methods. As such, the book is also suitable as a refresher for students entering a graduate control system program. In addition, it can also be used as an overview and a guide for practicing engineers to gain familiarity with the computer-aided design of classical control systems.

The material is organized like many of the introductory control system textbooks. The state-space modeling of systems is introduced early, in Chapter 4, to emphasize its importance in modeling real-world problems. However, for those following a syllabus without any state-space modeling content, Chapter 4 can be skipped without any loss of continuity.

This book is written based on the premise that the learning of control systems benefits from studying examples and then working problems with increasing levels of complexity. Each example is designed to illustrate a specific concept and usually contains a script of the MATLAB commands used for the model creation and the computation. Some examples are followed by "what-if" questions that allow the reader an immediate opportunity to answer related questions to appreciate some of the more intricate parts of the concept. Then several reinforcement problems are provided to further apply the technique to textbook-type problems. At the end of each chapter, one or two exploratory problems are posed to practice the techniques on user-defined systems. From Chapter 2 on, several comprehensive problems are presented that involve the analysis and/or design of real-world systems whose models are presented in Appendix A. To solve them, the reader is required to apply all the concepts discussed in the chapter.

To make effective use of this book, the user is expected to have some familiarity with MATLAB, including data entry, plotting, and simple computations. The MATLAB M-files are available at the BookWare Companion Resource Center, online at *http://www.brookscole.com/engineering/ee/bookware.html*. The files can be copied to your computer and used to solve all the examples, reinforcement problems, and comprehensive problems contained in the book. M-files that allow the user to do the exploratory problems are also given. The M-files also contain several special RPI functions that perform computations not found in either MATLAB or the Control System Toolbox.

ACKNOWLEDGMENTS

We would like to thank Tom Robbins, formerly of the PWS Publishing Company, for the opportunity to be part of the Bookware Companion Series and for his constant encouragement during the preparation of the original version of this book. The series editor, Bob Strum of the Naval Postgraduate School, provided helpful comments on its organization and interactive software. The following reviewers of the original manuscript provided inputs that were helpful in establishing the scope and orientation of the book:

Steve Adams, Swampscott, MA

Derek Atherton, University of Sussex

William Durfee, MIT

John A. Fleming, Texas A&M University

Jin Jiang, The University of Western Ontario

Mark Nagurka, Marquette University

Stephen Phillips, Case Western Reserve University

Walter Schaufelberger, Didaktikzentrum ETH

Some of the code in the RPI functions was developed by Jim O'Donnell during his time as a teaching assistant at Rensselaer. Catherine Yadlon, Hades Shragai, and Shaopeng Wang edited and checked the MATLAB M-files, and Craig Borghesani applied his MATLAB expertise to enhance the M-files and ensure their proper functioning. Walter Schaufelberger of ETH suggested the gain-sweep method that we employ in Chapters 8 and 9.

We appreciate the assistance of Theresa Buffington, Cheryl Graves, and Michelle Thompson of The Beacon Group, who made the revisions to the LaTeX and figure files, and Greg Hubit of Greg Hubit Bookworks, who coordinated the production process. At Brooks/Cole Publishing Company, we acknowledge the help of Bill Stenquist, Shelley Gesicki, Mary Vezilich, Marlene Thom, and Vernon Boes.

We are grateful to our respective employers (Unified Technologies, Inc. and Rensselaer Polytechnic Institute) for the use of facilities and software that was essential to the preparation of this book. Special thanks go to David Kassover for his able assistance with the installation of the LaTeX software used for its development.

Finally, we would like to thank our families for their support during the long hours we spent in preparing the text and the M-files.

D. K. Frederick
J. H. Chow
Troy, New York
July 1999

Introduction

PREVIEW

The book contains examples and a variety of problems related to the analysis and design of feedback control systems that can be solved with MATLAB® and the Control System Toolbox.[1] As such, it serves as a companion to conventional control systems textbooks. The examples and problems in this book are designed to reinforce the theory presented in these books. Computer files that will solve the examples and the problems in the book can be found on the Brooks/Cole web site. This chapter describes the organization of the topics in the book and discusses the ways to use the book and its computer files.

MATLAB AND THE CONTROL SYSTEM TOOLBOX

There are several high-level interactive control system design software packages that could have been used for this book. However, the combination of MATLAB and the Control System Toolbox is probably the most widely available package of this type in use in the academic environment and is certainly well suited to the task. MATLAB, which is derived from MATrix

[1]MATLAB is a registered trademark of The MathWorks, Inc.

LABoratory, provides the underlying computational and graphical tools for handling real and complex scalars, vectors, and matrices. The Control System Toolbox is a set of functions written in the MATLAB language that makes it convenient to build the models and perform the analysis that is commonly done in control systems engineering. For example, with the commands of the Control System Toolbox, one can easily create system models in state-space or transfer-function form and perform a variety of time-domain and frequency response types of analysis, with graphical output when appropriate. These multifaceted model-building and analysis tools can then be used as the basis for a wide range of control system design methods. For the purpose of this book, we have implemented several additional functions for control system analysis, which will be called RPI functions (for Rensselaer Polytechnic Institute). At the end of each chapter, we will provide a list of MATLAB functions used in that chapter, together with a brief synopsis of each function and its origin.

We assume that the user has some experience with MATLAB, including its basic linear algebra features and plotting commands, and do not attempt to explain its features and use. An excellent source for someone who has little or no experience with MATLAB is *The Student Edition of MATLAB* by Hanselman and Littlefield (1997).[2]

If the reader knows MATLAB reasonably well, there should not be any difficulty in using the commands of the Control System Toolbox as they are described and illustrated in the text, at least in their basic form. MATLAB and the Control System Toolbox have an excellent on-line help facility so the user can get a brief statement of the syntax of a function directly from the computer. The same on-line help facility is available for the RPI functions, which also serves as the documentation of these functions.

CROSS-REFERENCE OF TOPICS

Since this book is intended as a computer-aided design problems book for a first course in control engineering for continuous-time systems, we have used the models in the Laplace-transform transfer-function form for most of the examples and problems. However, state-space models and design methods are contained in Chapters 4 and 10, respectively.

The first part of the book, Chapters 2, 3 and 4, is on the building of models. The treatment includes both transfer-function and state-space models. The second part, Chapters 5, 6 and 7, introduces analysis techniques and performance measures in the time- and frequency-domain. The third and last

[2]See the bibliography at the end of the book for a detailed citation.

part, Chapters 8, 9 and 10, is on control design techniques. The topics are covered in many standard control engineering textbooks for continuous-time systems. The reader can use Table 1.1 to cross-reference the topics in this book with those found in a number of these control engineering textbooks. Detailed citations of the books in Table 1.1 are given in the bibliography.

WAYS TO USE THIS BOOK

Each chapter covers an important area of control-system analysis or design. Each section in a chapter deals with a key topic that is illustrated by one or more examples in which MATLAB is used. The important commands are often given in the solutions to the examples as "script" files. In MATLAB, these script files are called the M-files, since they have the extension `.m`. In following the discussion of the example, the reader can type the MATLAB commands in the script files in an interactive manner or use the M-files

TABLE 1.1 *Chapters in control-engineering textbooks corresponding to the chapters of this book*

Chapter and Topic	*D'Azzo & Houpis*	*Dorf & Bishop*	*Franklin, Powell, & Emami-Naeini*
2. Single-block systems	5	2	3
3. Multi-block systems	5	2,4	3
4. State-space models	3	3	7
5. Root locus	7	7	5
6. Frequency response	8	8,9	6
7. System performance	3,6	4,5	4
8. PID design	10	12	4
9. Lead-lag design	11	10	6
10. State-space design	13	11	7

Chapter and Topic	*Kuo*	*Nise*	*Ogata*	*Phillips & Harbor*	*Rohrs, Melsa, & Schultz*
2. Single-block systems	3	2	1,4	2	2
3. Multi-block systems	3	2	1	2	2
4. State-space models	5	3	3	3	3
5. Root locus	8	8	6	7	7
6. Frequency response	9,10	10	8	8	5
7. System performance	7	7	4,5	4,5	6
8. PID design	10	9	10	9	8
9. Lead-lag design	9,10	11	9	9	8
10. State-space design	10	12	11,12	10	3

that can be downloaded from the Brooks/Cole web site. A description of the M-files is provided below. To save space, most of the plotting commands are not listed in the script file in the example solution. However, they can be found in the M-files.

Following each example, one or more what-if's may be posed to examine the effect of the variations of the key parameters. Each example is followed by several reinforcement problems that are similar in scope and solution approach. Comments to alert the reader to the special features of the problems are included in order to enhance the learning experience. Partial answers to these problems are given at the end of the chapter.

After all the topics in a chapter have been covered, one or more exploratory problems may be posed to allow the student to apply the MATLAB functions discussed in the chapter to models of any kind. Then several comprehensive problems based on models of physical systems are included. The models of the physical systems, together with the descriptions of the systems, are given in Appendix A. These comprehensive problems tend to exercise the concepts and the MATLAB commands introduced in the chapter to give the reader a more complete picture of the overall control analysis and design process.

The software obtainable from the web site will be of considerable interest to the reader because it contains the MATLAB M-files that will run each of the examples and problems. These files are in the MS-DOS® format and can be copied directly onto a PC.[3] The MS-DOS files can also be read into a Macintosh®- or a UNIX®-based workstation using the proper file-transfer facilities.[4,5] These M-files can be run on the corresponding computer platform that has the full MATLAB version 5.0 and version 4.1 of the Control System Toolbox. The M-files for the RPI functions are also available on the web site.

After studying a new topic in a conventional text on control systems engineering, the reader should study the corresponding section in this book. Then the M-files containing the commands that solve the related examples and problems should be copied from the web site and run in MATLAB by typing its filename (without the `.m` extension). The convention for the example and problem M-files is as follows.

For each chapter, a single M-file runs all the examples. For example, the M-file for the examples in Chapter 2 is `ch2ex.m`.[6] Each example file starts with a menu from which the user can choose any of the examples in the corresponding chapter. In particular, the user can run these example M-files to generate all the MATLAB plots in this book. The M-files for the what-if

[3]MS-DOS is a registered trademark of Microsoft Corporation.

[4]Macintosh is a registered trademark of Apple Computer, Inc.

[5]UNIX is a registered trademark of American Telephone and Telegraph Company.

[6]In the text, MATLAB commands and M-files are printed in this type style.

questions start with the prefix `wif`, the reinforcement problems with `p`, and the comprehensive problems with `cp`. For example, the file `wif6_4b.m` contains the MATLAB commands to solve part b of the what-if question associated with Example 6.4, and the file `p2_4.m` solves Reinforcement Problem 2.4 in Chapter 2. The instructions to run the M-files for the exploratory problems are given with the problem statement.

To see the MATLAB commands used in an M-file, the `echo` command can be entered before the execution of the file. You may find it helpful to use the `diary` command to capture the contents of the MATLAB command window as the M-file executes. The resulting diary file can be printed and carefully studied afterwards.

In the M-files, we have included `pause` commands so that the display of computational results will not scroll off the computer display screen before the user has a chance to view them. When the program execution pauses, the characters

```
******>
```

are displayed. Simply press any key to continue. When the symbol

```
K>>
```

is displayed, the program is in the `keyboard` mode. This is implemented to allow the user to examine the newly-created plot and, if desired, make a hardcopy of the plot by using the `print` command. The user can also enter any MATLAB commands at this point. To continue the execution of the program, simply enter the command `return`.

For the user who wants to apply the tools discussed in this book to other systems, one convenient way is to use a copy of an appropriate M-file as a template. The system model in the template can be replaced with the new data, and additional MATLAB and Control System Toolbox commands can be inserted if desired. Then the modified file can be run by typing its name without the `.m` extension. In this way, the user can quickly investigate the effects of changing the values of model parameters, controller gains, and inputs on whatever dynamic system is under consideration. When used properly, the combination of MATLAB and the many M-files included with this book can greatly accelerate the process of understanding the subject of control engineering.

Single-block Models and Their Responses

CHAPTER 2

PREVIEW

The starting point of a control system design is the formulation of a system model. In a similar manner, Control System Toolbox functions operate on the system model. The models of a signal and the transfer function of a fixed linear system can be represented in several different mathematical forms. One of these forms is obtained by applying the Laplace transform to the system, resulting in its *transfer function*, expressed as a ratio of polynomials in *s*. A second method for describing a system in Matlab is in terms of the zeros, poles, and gain of its transfer function. In this chapter, we show how both of these methods can be used to implement the model of a single subsystem, as would be represented by a single block in a block diagram. The analysis for a system with multiple blocks is considered in the next chapter, and a third form of representing a fixed linear system, namely the state-space form, will be discussed in Chapter 4.

TRANSFER FUNCTIONS

Consider a fixed single-input/single-output linear system with input $u(t)$ and output $y(t)$ given by the differential equation

$$\ddot{y} + 6\dot{y} + 5y = 4\dot{u} + 3u \tag{2.1}$$

Applying the Laplace transform to both sides of (2.1) with zero initial conditions, we obtain the transfer function of the system from the input $U(s)$ to the output $Y(s)$ in TF form as the ratio of polynomials[1]

$$G(s) = \frac{4s + 3}{s^2 + 6s + 5} \tag{2.2}$$

Alternatively, this transfer function can be expressed in terms of its zeros z_i, poles p_j, and gain K in the factored, or ZPK form, as

$$G(s) = \frac{4(s + 0.75)}{(s + 1)(s + 5)} \tag{2.3}$$

which shows that $G(s)$ has a single zero at $s = -0.75$, two poles at $s = -1$ and -5, and a gain of 4.

If we know the numerator and denominator polynomials of $G(s)$, we can represent the model in MATLAB as a Linear Time-invariant (LTI) object in Transfer Function (TF) form by (i) creating a pair of *row* vectors containing the coefficients of the powers of s, in descending order, of the numerator and denominator polynomials, and (ii) using the `tf` command to create the TF object. For the transfer function in (2.2), we could enter `numG = [4 3]` and `denG = [1 6 5]` to define the polynomials, followed by `G1 = tf(numG, denG)` to create the LTI object. Assuming that we do not need to have the numerator and denominal polynomials available in the MATLAB workspace, we could build the TF object with the single statement `G1 = tf([4 3], [1 6 5])`.

If the transfer function is known in terms of its zeros, poles, and gain, we can create the model as a LTI object in ZPK form by (i) entering *column* vectors for the zeros and poles, and enter the gain as a scalar and (ii) using the `zpk` command to create the ZPK object. To create the system described in (2.3), we could enter the commands `zG = -0.75`, `pG = [-1; -5]`, `kG = 4`, and `G2 = zpk(zG,pG,kG)`. Or, we could use the single command `G2 = zpk(-0.75,[-1; -5],4)`.

When a model has been described in MATLAB in either one of these forms, the Control System Toolbox provides the ability to convert from one form to the other. For example, if the system model has been created in TF form in terms of its numerator and denominator polynomials as `Gtf`, we can create the ZPK form merely by entering `Gzpk = zpk(Gtf)`. Then, to determine the zeros, poles, and gain of $G(s)$, we can enter `[zz,pp,kk] = zpkdata(Gzpk,'v')`. In similar fashion, we could build the system in ZPK form as, say `Sxx`, convert it to TF form via `Svv = tf(Sxx)`, and then enter `[nn,dd] = tfdata(Svv,'v')`.

[1] $U(s)$ is the Laplace transform of the variable $u(t)$, i.e., $U(s) = \mathcal{L}[u(t)]$.

There are two things to note about these commands. First, the name assigned to the TF or ZPK system is arbitrary, as evidenced by the use of `G1, G2, Gzpk, Gtf, Sxx` and `Svv` as LTI-object names. Second, the use of the string `'v'` (the single quotes around the letter `v` make it a string in MATLAB) as the second argument in the `tfdata` and `zpkdata` commands forces the outputs to be written as vectors. For example, once the TF object `G1` has been created as shown above, the zeros, poles, and gain of $G(s)$ can be obtained by the command `[z,p,k] = zpkdata(G1,'v')` to extract the desired quantities into the MATLAB workspace.

Once we have the system model expressed as an LTI object, in either the TF or the ZPK form, we can obtain a graphical representation of the transfer function's poles and zeros by using the `pzmap` command from the Control System Toolbox. Assuming the name of the object is `G`, the command `pzmap(G)` will display an appropriate region of the s-plane with the zeros indicated by the symbol "○" and the poles by the symbol "×".

EXAMPLE 2.1
Build $G(s)$ as a TF Object

Create a TF object for the fourth-order system whose differential equation is

$$y^{(iv)} + 10\dddot{y} + 30\ddot{y} + 40\dot{y} + 24y = 4\ddot{u} + 36\dot{u} + 32u \tag{2.4}$$

where the symbol $y^{(iv)}$ denotes the fourth derivative of $y(t)$ with respect to time. Then display the object's properties and extract the numerator and denominator polynomials from the TF object and display them as row vectors. Draw a pole-zero plot.

Solution

Taking the Laplace transform of the differential equation with zero initial conditions, we obtain the polynomial form of the transfer function as

$$G(s) = \frac{4s^2 + 36s + 32}{s^4 + 10s^3 + 30s^2 + 40s + 24}$$

The MATLAB commands in Script 2.1 will create $G(s)$ as a TF object, display all of its current properties (at the moment we are interested in only the first two), create the two row vectors `nn` and `dd` that contain the coefficients of the numerator and denominator polynomials, and draw the pole-zero plot shown in Figure 2.1.

MATLAB Script

```
% Script 2.1:  Create G(s) as a TF object
numG = [4 36 32]                        % enter transfer function numerator
denG = [1 10 30 40 24]                  % ...and denominator polynomials
G = tf(numG,denG)                       % Create G(s) as a TF object
get(G)                                  % Show properties of TF object
% Extract numerator & denominator polynomials from TF object
[nn,dd] = tfdata(G,'v')
pzmap(G)                                % draw pole-zero plot from TG object
```

CHAPTER 2

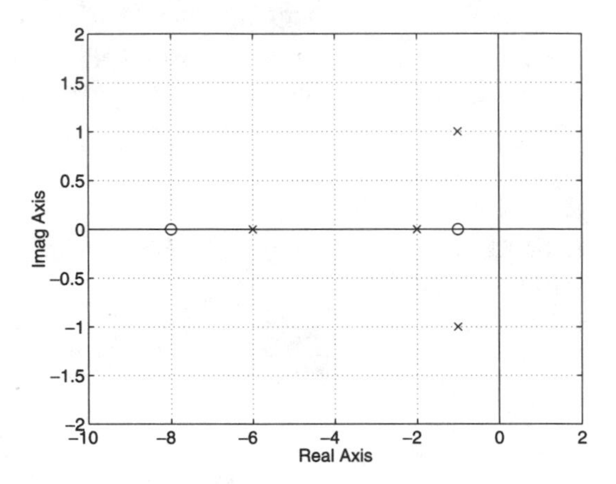

FIGURE 2.1 *Pole-zero plot for the system in Example 2.1*

EXAMPLE 2.2 ***Create G(s) as a ZPK object***

Use the TF object created in Example 2.1 to obtain the model of the system described by (2.4) as a ZPK object. Then extract its zeros, poles, and gain and show that they are the same as those of the TF object created in the prior example.

Solution

We repeat the initial steps of Example 2.1 to create the TF object `G`. Then we convert it to a ZPK object named `GG` with the command `GG = zpk(G)`. To compare the zeros, poles, and gain of the two objects, we use the `zpkdata` command on both objects, but with different output arguments so the results have distinct names. Doing this, we find that both systems have zeros at $s = -1.0$ and -8.0, poles at $s = -1.0 \pm j1.0$, -2.0, and -6.0, and a gain of 4.0.

MATLAB Script

```
% Script 2.2:  Convert from TF to ZPK form
n = [4 36 32]                    % enter transfer-function numerator
d = [1 10 30 40 24]              % ...and denominator polynomials
G = tf(n,d)                      % Create G(s) as a TF object
GG = zpk(G)                      % Convert G(s) into ZPK object
[zz,pp,kk] = zpkdata(GG,'v')     % Extract poles and zeros from
.                                % .....ZPK form into arrays
[z,p,k] = zpkdata(G,'v')         % Extract poles and zeros from
                                 % ....TF form into arrays
```


REINFORCEMENT PROBLEMS

For the system whose transfer function is given below as a ratio of polynomials, implement the model in TF form, convert it to ZPK form, and draw a pole-zero plot.

P2.1 Third-order system.

$$G(s) = \frac{2s^2 + 3s + 1}{s^3 + 2.4s^2 + 1.8s + 2}$$

P2.2 Fourth-order system.

$$G(s) = \frac{3s^2 + 6s + 4}{s^4 + 3s^3 + 8s^2 + 4s + 2}$$

For the system whose zeros, poles, and gain are given below, implement the model in ZPK form, convert it to TF form, and draw a pole-zero plot.

P2.3 Fourth-order system with complex poles. $G(s)$ has zeros at $s = -2$ and -4, poles at $s = -1, -4 \pm j6, -20$, and a gain of 150.

P2.4 Fourth-order system with real poles. $G(s)$ has zeros at $s = -3$ and $-8 \pm j2$, poles at $s = -1, -4, -10, -30$, and a gain of 5.

RESIDUES AND IMPULSE RESPONSE

Once a transform (or transfer function) $G(s)$ has been defined in MATLAB, operations can be performed to compute and display the corresponding time function $g(t)$, known as the *inverse Laplace transform*. If the poles are known, $G(s)$ can be written as a partial-fraction expansion from which the individual transform terms can be associated with time functions. MATLAB provides the function `residue` to compute the poles and their corresponding residues in a partial-fraction expansion, from which we can obtain $g(t)$ in analytical form. Alternatively, using the property that $g(t)$ is the unit-impulse response corresponding to $G(s)$, the Control System Toolbox function `impulse` can be directly applied to obtain $g(t)$ in numerical form, from which it can be plotted.

CHAPTER 2

EXAMPLE 2.3
Find the Partial-Fraction Expansion and $g(t)$

The transfer function of a fixed linear system is

$$G(s) = \frac{3s + 2}{2s^3 + 4s^2 + 5s + 1} \tag{2.5}$$

Create the transfer function in MATLAB and determine its poles and zeros. Perform a partial-fraction expansion of $G(s)$, and plot the impulse response of the system.

Solution

The system is implemented as a TF object by defining the row vectors `numG = [3 2]` and `denG = [2 4 5 1]`, which contain the coefficients of the numerator and denominator polynomials, respectively, and using them as arguments of the `tf` command, as indicated in Script 2.3. To obtain the zeros, poles, and gain of $G(s)$, we enter `[zG,pG,kG] = zpkdata(G,'v')`. Doing this, we find that $G(s)$ has a zero at $s = -0.6667$ and poles at $s = -0.2408$ and $-0.8796 \pm j1.1414$, and the gain is $3/2 = 1.50$. Thus we can write the transfer function in factored form as

$$G(s) = \frac{1.5(s + 0.6667)}{(s + 0.2408)(s + 0.8796 - j1.1414)(s + 0.8796 + j1.1414)}$$

To write $G(s)$ in a partial-fraction expansion, we use the `residue` command as in `[resG,polG,otherG] = residue(numG,denG)`. Note that the `residue` command does not accept LTI objects as its input argument. Instead, it requires the numerator and denominator coefficient vectors `numG` and `denG` as input arguments.

The result is a column vector `resG` containing the residue at each of the poles of $G(s)$, a column vector `polG` containing the poles, and a scalar `otherG` consisting of a constant that will be nonempty only if the degree of the numerator of $G(s)$ equals the degree of the denominator. Doing this, we obtain the poles and residues given in Table 2.1.

TABLE 2.1 *Poles and zeros of $G(s)$ in Example 2.3*

pole	*residue*
-0.2408	0.3734
$-0.8796 + j1.1414$	$-0.1867 - j0.5526$
$-0.8796 - j1.1414$	$-0.1867 + j0.5526$

The third result produced by the `residue` command is the scalar `otherG`, which has no value because there are no other terms in the expansion. MATLAB sets it to `[ ]`, which is referred to as the *empty element*.

From these MATLAB results we can write the partial-fraction expansion of the transfer function as

$$G(s) = \frac{0.3734}{s + 0.2408} + \frac{-0.1867 - j0.5526}{s + 0.8796 - j1.1414} + \frac{-0.1867 + j0.5526}{s + 0.8796 + j1.1414} \tag{2.6}$$

The impulse response can be computed by using the `impulse` command, which can take one of several different forms. The simplest of these is to enter `impulse(G)`, which will cause a plot of $g(t)$ to be displayed using a time interval selected by MATLAB. Another option is to enter `impulse(G,time)` where `time` is a *column* vector of time points that has been defined previously. A third option is to use a single left-hand argument as in `y = impulse(G,time)` with the time vector specified as before. This form of the command will return the output values for each time point in the column vector `y`, which can then be plotted versus time by entering `plot(time,y)`. The corresponding impulse response appears in Figure 2.2.

MATLAB Script

```
% Script 2.3:  Partial-fraction expansion and impulse response
numG = [3 2], denG = [2 4 5 1]          % numerator & denominator of G(s)
G = tf(numG,denG)                        % Create G(s) as TF object
[zG,pG,kG] = zpkdata(G,'v')             % zeros, poles, & gain of G(s)
[resG,polG,otherG] = residue(numG,denG) % do PFE to get residues
impulse(G)                               % impulse response
```

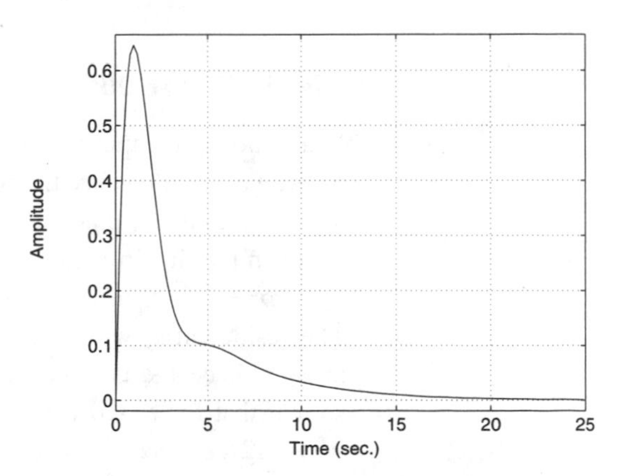

FIGURE 2.2 *Impulse response for Example 2.3*

Comment: If the degree of the numerator of $G(s)$ is the same as the degree of its denominator, the `otherG` part of the result of the `residue` command will be a constant which represents an impulse component in $g(t)$. While the `residue` command returns this value correctly, the `impulse` command will not include it in the response. It is up to the user to keep track of any nonempty `otherG` part of a partial-fraction expansion.

WHAT IF? Try changing the coefficients of the numerator of $G(s)$ while leaving the denominator alone. Look at the effects on the poles, zeros, gain, residues, and impulse response. You should find that all of the above are affected except the poles. Looked at another way, by changing the numerator of $G(s)$ but not its denominator, you are affecting the weighting of the system's mode functions but not the mode functions themselves. You can use the file `wif2_3.m` to get started. ■

REINFORCEMENT PROBLEMS

For each of the differential equations that follows, create the system model as a TF object. Then use MATLAB to find the zeros, poles, and gain of $G(s)$; to perform a partial-fraction expansion; and to compute and plot the unit-impulse response.

P2.5 Third-order system.

$$\dddot{y} + 7\ddot{y} + 32\dot{y} + 60y = 2\ddot{u} + 11\dot{u} + 5u$$

P2.6 Fourth-order system.

$$2y^{(iv)} + 7\dddot{y} + 11\ddot{y} + 12\dot{y} + 4y = 3\ddot{u} + 4\dot{u} + 5u$$

TIME RESPONSE DUE TO DISTINCT POLES

In addition to computing the time response of the output, we can use MATLAB to evaluate and display the time response due to an individual real pole or to a pair of complex poles. Recall that a real pole at $s = p$ with residue r will result in the time function $y(t) = r\epsilon^{pt}$ for $t > 0$. If a pole is complex, say at $s_1 = \sigma + j\omega$, then its complex conjugate $s_2 = \sigma - j\omega$ will also be a pole. The residue at s_1 will be a complex number, say $r_1 = K\epsilon^{j\phi}$, and the residue at s_2 will be the complex conjugate of r_1, namely, $r_2 = K\epsilon^{-j\phi}$. It can be shown that for $t > 0$ the time response due to this pair of complex poles is $y(t) = 2K\epsilon^{\sigma t}\cos(\omega t + \phi)$.

To simplify the calculation of these responses, we have created two functions designated as "RPI functions" named `rpole2t` and `cpole2t`. The former has as its arguments the value of the real pole, its residue, and a time vector. The function returns a column vector of time response values, one per time point. The latter function has the value of the complex pole with the *positive* imaginary part as its first argument, the residue at this pole as the second argument, and a time vector as the final argument.

EXAMPLE 2.4
Responses Due to Individual Poles

Use the poles and residues for the transfer function $G(s)$ in Example 2.3 to display the components of $g(t)$ due to the real pole at $s = -0.2408$ and the complex poles at $s = -0.8796 \pm j1.1414$. Verify that the sum of these two responses equals the impulse response shown in Figure 2.2.

Solution

Referring to Table 2.1, we use the RPI function `rpole2t` with the pole $s = -0.2408$ and the residue $r = 0.3734$ to obtain the response due to the one real pole of $G(s)$. The resulting time function is the inverse Laplace transform of the first term in (2.6). For the second and third terms in (2.6), which result from the complex poles, we use the pole having the positive imaginary part, namely $s = -0.8796 + j1.1414$ and its residue $r = -0.1867 - j0.5526$ as arguments of the RPI function `cpole2t`. The MATLAB instructions that will accomplish these steps and plot the responses are shown in Script 2.4, and the resulting responses are shown in Figure 2.3.

MATLAB Script

```
% Script 2.4: Responses due to individual poles
numG = [3 2], denG = [2 4 5 1]           % create G(s) as ratio of polys
[resG,polG,otherG] = residue(numG,denG) % residues & poles of G(s)
t = [0:0.1:20]';                         % column vector of time points
ycmplx = cpole2t(polG(1),resG(1),t);     % response due to complex poles
yreal = rpole2t(polG(3),resG(3),t);      % response due to real pole
ytot = ycmplx + yreal;                   % y(t) is sum of the two
plot(t,ytot,t,ycmplx,t,yreal,'--')       % plot the three curves
```

CROSS-CHECK Verify that the plot of `ytot` labeled "total" in Figure 2.3, which is the sum of the two individual responses, agrees with the response shown in Figure 2.2 that was obtained by using the `impulse` command. ■

FIGURE 2.3 *Individual responses for Example 2.4*

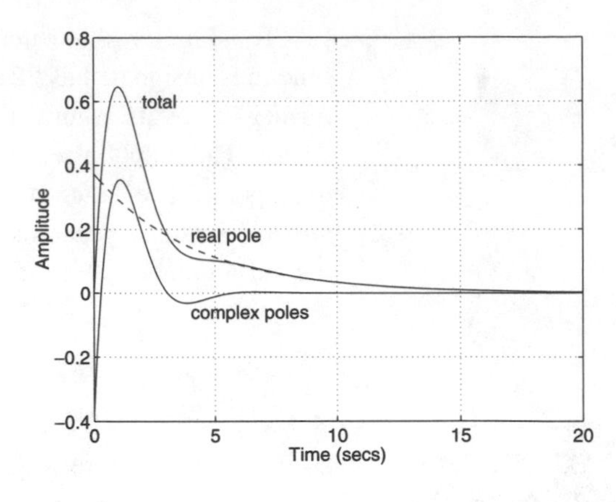

REINFORCEMENT PROBLEMS

P2.7 Third-order system with responses due to individual poles. For the transfer function $G(s)$ used in Problem 2.5, compute and plot the individual responses due to the real and complex poles. Verify that the sum of these responses agrees with the impulse response $g(t)$ from Problem 2.5.

P2.8 Fourth-order system with responses due to individual poles. Repeat the steps of Problem 2.7 as applied to the transfer function $G(s)$ in Problem 2.6.

TIME RESPONSE DUE TO REPEATED POLES

Up to this point the discussion has been restricted to distinct poles, either real or complex. For a repeated pole there will be more than one term in the time response, with the number of terms depending on the multiplicity of the pole. For a pole at $s = p$ of multiplicity $m = 2$, we can write the Laplace transform of the response as

$$G(s) = \frac{A}{(s-p)^2} + \frac{B}{s-p} + \text{terms involving other poles} \tag{2.7}$$

where

$$A = [(s-p)^2 G(s)]\,\big|_{s=p} \quad \text{and} \quad B = \left\{\frac{d}{ds}[(s-p)^2 G(s)]\right\}\bigg|_{s=p} \tag{2.8}$$

For $t > 0$, the corresponding impulse response is

$$g(t) = At\epsilon^{pt} + B\epsilon^{pt} + \text{terms involving other poles} \tag{2.9}$$

Similar relationships exist for poles of multiplicity three or higher and for repeated complex poles, but they are seldom needed and will not be covered here. The residue command is able to compute the coefficients for repeated poles, and for $m = 2$ they are listed as B followed by A.

EXAMPLE 2.5
Repeated Poles

For the system whose differential equation is

$$\ddot{y} + \dot{y} + 0.25y = \dot{u} + 2u$$

do a partial-fraction expansion and write the impulse response as the sum of two individual functions of time.

Solution

The transfer function is

$$G(s) = \frac{s+2}{s^2 + s + 0.25} = \frac{s+2}{(s+0.5)^2} = \frac{A}{(s+0.5)^2} + \frac{B}{s+0.5}$$

which has a pole at $s = -0.5$ of multiplicity two. Using the residue command results in the column vectors

```
resG = 1.0000   and   polG = -0.5000
       1.5000                -0.5000
```

and the scalar otherG = [ ]. Associating the first element of resG with B and the second element with A, we can write

$$G(s) = \frac{1.5}{(s+0.5)^2} + \frac{1.0}{s+0.5}$$

From (2.7) and (2.9), where there are no other poles, we see that for $t > 0$ the impulse response is

$$g(t) = 1.5t\epsilon^{-0.5t} + 1.0\epsilon^{-0.5t}$$

Note the use in the script of the .* operator in the expression for g2. This causes element-by-element multiplication of the vector t and the result of the rpole2t function, which is the exponential term $1.5\epsilon^{-0.5t}$.

MATLAB Script

```
% Script 2.5:  Response due to repeated poles
numG = [1 2], denG = [1 1 0.25]          % num & den polys of G(s)
[resG,polG,otherG] = residue(numG,denG)  % residues & poles of G(s)
t = [0:0.1:20]';                         % column vector of time points
g1 = rpole2t(polG(1),resG(1),t)          % purely exponential response term
g2 = t.*rpole2t(polG(2),resG(2),t)       % time * exponential response term
plot(t,g1,t,g2);grid                     % plot the two terms separately
```

REINFORCEMENT PROBLEMS

Use MATLAB to obtain a partial-fraction expansion of the system's transfer function and to write the impulse response as a sum of individual time functions. Also plot the impulse response.

P2.9 Third-order system with one repeated pole.

$$\dddot{y} + 3.75\ddot{y} + 4.5\dot{y} + 1.6875y = 6\dot{u} + 4u$$

P2.10 Fourth-order system with two repeated poles.

$$y^{(iv)} + 5\dddot{y} + 8.25\ddot{y} + 5\dot{y} + y = 3\ddot{u} + 15\dot{u} + 12u$$

STEP RESPONSE

The Laplace transform of a system's unit-step response is the product of the system's transfer function, $G(s)$, and $1/s$, the transform of the unit-step function. The poles of the resulting transform are the poles of $G(s)$ and a pole at $s = 0$ (due to the unit-step input). The zeros and gain of the step response are the same as those of the transfer function. The step response can be computed and plotted using the `step` command from the Control System Toolbox. This command has the same options as does the `impulse` command for plotting and returning numerical values.

The gain at $s = 0$ is $G(0)$ and is known as the *DC gain*. It is the ratio of the constant terms in the numerator and denominator polynomials. We can compute the DC gain directly from either the TF or ZPK object using the Control System Toolbox command `dcgain(G)`.

In the following example we use MATLAB to construct the Laplace transform of a step response, to plot the response with the `impulse` command, and to compare the result with a plot obtained using the `step` command. We also illustrate the use of the initial- and final-value theorems.

EXAMPLE 2.6
Step Response from G(s)

For the transfer function $G(s)$ used in Example 2.3 and given by (2.5), obtain a plot of the step response by adding a pole at $s = 0$ to $G(s)$ and using the `impulse` command to plot the inverse Laplace transform. Compare the response with that obtained with the `step` command applied to $G(s)$. Also determine the system's DC gain. Use the initial- and final-value theorems to check your results.

Solution Because the system's transfer function is the same as that in the previous example, we use the same values for `numG` and `denG`. To add a pole at $s = 0$, we multiply the denominator polynomial by s. In MATLAB this can be done in either of two ways. One method is to use the `conv` command, which will multiply two polynomials by performing a discrete convolution of the row vectors containing their coefficients. In this example we want to multiply the denominator of $G(s)$ by the polynomial s, which in MATLAB is represented by the row vector `[1 0]`. To do this, we enter `denstep = conv(denG,[1 0])`.

The other method is to append a `0` to the row vector `denG` with the statement `denstep = [denG 0]`. The numerator of the transform of the step response is the same as that of $G(s)$, so we write `numstep = numG`. Having obtained the transform of the unit step response in terms of the row vectors `numstep` and `denstep`, we create the TF object `Gstep` and use the `impulse` command to get the inverse transform, which is the step response shown in Figure 2.4. Alternatively, we could have used the TF object `G` constructed from the polynomials `numG` and `denG` and the `step` command to generate the same plot. The calculation of the DC gain gives a value of 2.0, which agrees with $G(0)$ as obtained by letting $s = 0$ in (2.5). These commands are contained in Script 2.6.

SINGLE-BLOCK MODELS AND THEIR RESPONSES

MATLAB Script

```
% Script 2.6: Obtain step response from G(s)
numG = [3 2], denG = [2 4 5 1]          % num & denom of G(s)
numstep = numG                          % rename numerator, no change
denstep = [denG 0]                      % add pole at s=0 to G(s)
Gstep = tf(numstep,denstep)             % create G(s)/s as TF object
impulse (Gstep,30)                      % use impulse with G(s)/s
G = tf(numG,denG)                       % create G(s) as TF object
step(G,30)                              % 30-sec step response with G(s)
DC_gain = dcgain(G)                     % calculate G(0)
```

If $y_s(t)$ denotes the system's unit step response, we can see from Figure 2.4 that $y_s(0+) = 0$ and $y_s(\infty) = 2.0$. To verify these values analytically, we write the Laplace transform of the step response as

$$Y_s(s) = \frac{1}{s}G(s) = \frac{3s + 2}{s(2s^3 + 4s^2 + 5s + 1)}$$

The initial-value theorem gives

$$y_s(0+) = \lim_{s\to\infty} sY_s(s) = \lim_{s\to\infty} \frac{3s + 2}{2s^3 + 4s^2 + 5s + 1} = 0$$

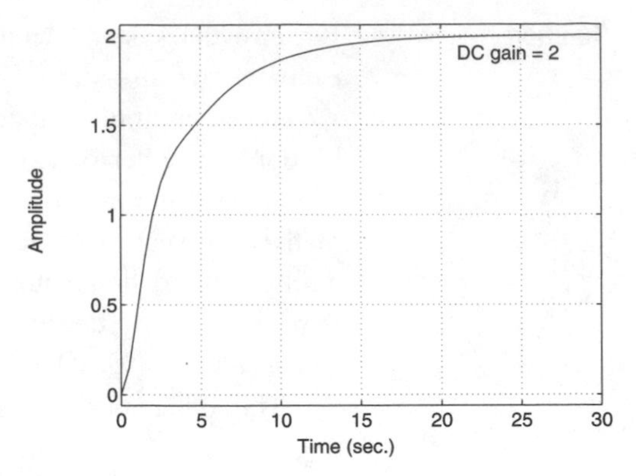

FIGURE 2.4 *Step response for Example 2.6*

and the final-value theorem gives

$$y_s(\infty) = \lim_{s \to 0} sY_s(s) = \lim_{s \to 0} \frac{3s + 2}{2s^3 + 4s^2 + 5s + 1} = 2$$

WHAT IF? Find and plot the response of the system in Example 2.3 to the unit ramp function by adding two poles at $s = 0$ to $G(s)$. ■

REINFORCEMENT PROBLEMS

P2.11 Step response. Obtain a plot of the unit-step response for the system described in Problem 2.5. Use the final-value theorem to verify the steady-state value of the response.

P2.12 Another step response. Repeat the steps of Problem 2.11 for the system described in Problem 2.6.

RESPONSE TO A GENERAL INPUT

In addition to computing and plotting the impulse and step responses of a system, MATLAB can be used to find and display the response to general functions of time. This is done with the `lsim` command, which can be used in a variety of ways. In its simplest form, the user specifies the system as an LTI object (in either TF or ZPK form), a vector of input values, and a vector of time points.

If the `lsim` command, which perfoms linear simulation, is given with no output variable, the plot of the response is drawn but no numerical values are returned. A more useful way to use this command is to specify the system output (`y`) as a result and then to plot both the output and the input (`u`) versus time. We illustrate the use of `lsim` in the following example by solving for the zero-state response to an input signal that is piecewise constant.

EXAMPLE 2.7
Response to Pulse Input

Find the zero-state response of the system discussed in Examples 2.3, 2.4, and 2.6 to the input

$$u(t) = \begin{cases} 0 & t < 0 \\ 2 & 0 \leq t < 2 \\ 0.5 & t \geq 2 \end{cases}$$

Solution

Once the system object has been defined, we need to create vectors of time points and input values. The only constraints are that the time points must be uniformly spaced and that the two vectors must have the same number of elements. We start by defining `time` to be a column vector that has a sufficiently small time increment (say, 0.02 s) to yield a smooth response plot and a large enough maximum time to show all of the transients. The command `time = [0:0.02:10]'` will accomplish this for the system and input signal under consideration.

The input vector `u` can be created in a number of ways. As shown in Script 2.7, we first make `u` be a vector having the same dimension as `time` but with all of its values equal to 2.0. Then we use a `for` loop to change the values of all the elements corresponding to $t \geq 2$ from 2.0 to 0.5. Note that the range of the index `ii` for which the values are changed is calculated by MATLAB based on the properties of `time` and `u` by using the commands `min`, `find`, and `length`. Specifically, `min(find(time>=2.0))` is the index of the first time point that is greater or equal to 2.0 s. This is because `find(time>=2.0)` returns the indices of *all* the time points that are 2.0 or greater, and the `min` function selects the smallest of these. Also, `length(u)` is the index of the last element of the input vector `u`. This approach has the advantages that one does not have to calculate the starting and ending values of the index `ii` and that no modifications are required if the number of time points is subsequently changed, due to a change in either the time increment or the maximum time.

The values of the response are computed by the command `lsim` and are stored in the column vector `y`. Then both `y` and `u` are plotted versus `time`, resulting in Figure 2.5.

MATLAB Script

```
% Script 2.7:  Response to piecewise constant input
G = tf([3 2],[2 4 5 1])                    % build G(s) as TF object
time = [0:0.02:10]';                       % 501 time points, every 0.02 s
u = 2.0*(1+0*(time))                       % vector of 2's with length of "time"
for ii = ...                               % continuation of command on next line
  min(find(time>=2.0)):length(u),
  u(ii) = 0.5;                             % change values from 2.0 to 0.5
end
y = lsim(G,u,time)                         % compute response
plot(time,y,time,u,'--')                   % use dashed line for input
text(3.5,1.7,'output')                     % put labels for curves directly on plot
text(4.0,0.58,'input')
```

FIGURE 2.5 *Input and response of the system in Example 2.7*

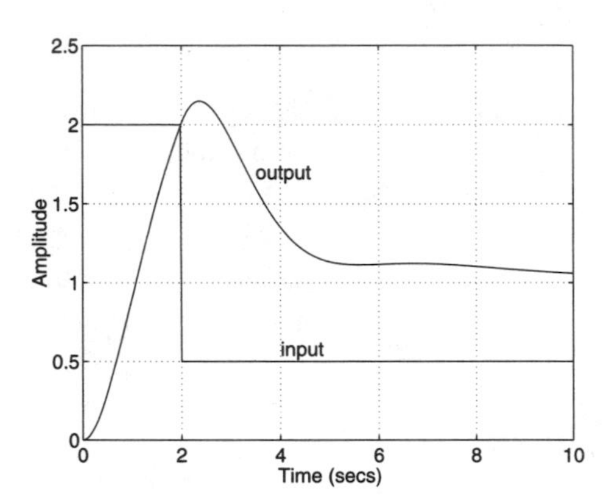

WHAT IF? Suppose the input to the system is that used in the example but is delayed by 1 second. Modify the commands in Script 2.7 to compute and plot the response. ■

REINFORCEMENT PROBLEMS

P2.13 Response to ramp-to-constant input. Compute and plot the zero-state response of the system described in Problem 2.5 to the input $u_1(t)$ shown in Figure 2.6.

P2.14 Response to one cycle of a sine wave. Compute and plot the zero-state response of the system described in Problem 2.6 to the input $u_2(t)$ shown in Figure 2.6.

FIGURE 2.6 *Input signals for Problems 2.13 and 2.14*

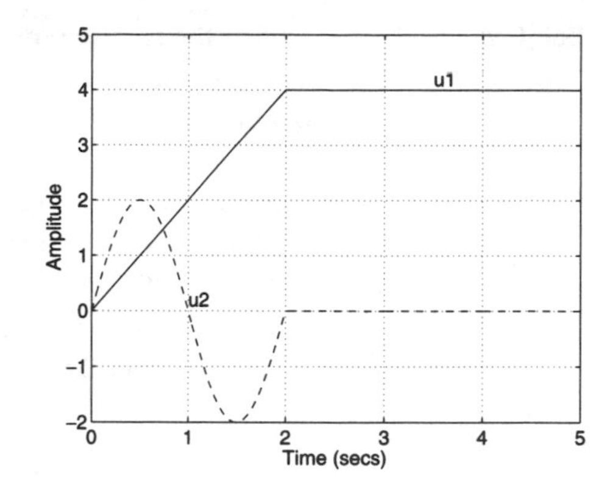

POLES AND STABILITY

If p_i is a pole of $G(s)$, then the natural, or zero-input, response of $G(s)$ will consist of the *mode functions* $\epsilon^{p_i t}$ if p_i is distinct, and $t^q \epsilon^{p_i t}$, $q = 0, 1, \ldots, r - 1$, if p_i has multiplicity r. Thus the natural response will decay to zero if $\text{Re}[p_i] < 0$ for $i = 1, \ldots, n$—that is, if all the poles are in the open left-half of the s-plane, i.e., the left half-plane excluding the imaginary axis. Such a system is said to be *asymptotically stable*. If all the poles are in the open left-half plane except for distinct poles on the imaginary axis and at the origin, the natural response will consist of undamped sinusoids or a nonzero constant, and the system is said to be *marginally stable*. If some of the poles are in the right-half of the s-plane or on the imaginary axis with multiplicity greater than one, then the natural response will be unbounded and the system is said to be *unstable*.

The following example illustrates two ways to determine the system's stability: (i) use the `roots` command on the denominator of $G(s)$, or (ii) use the `pole` command on the LTI object `G`.

EXAMPLE 2.8 *Poles and System Stability*

Find the poles of the transfer function

$$G(s) = \frac{1.5s + 1}{s^3 + 2s^2 + 2.5s + 0.5}$$

and determine whether the system is stable. Plot the poles and zeros of $G(s)$ in the s-plane. Finally, demonstrate the system stability by simulating the impulse response.

Solution Define the row vectors `numG = [1.5 1]` and `denG = [1 2 2.5 0.5]` to represent the numerator and denominator of $G(s)$, respectively. Using the command `roots(denG)`, we find the poles of $G(s)$ to be at $s = -0.2408$ and $-0.8796 \pm j1.1414$. Alternatively, we can build the TF object `G` and use it as the input argument of the `pole` command. Since all the poles have negative real parts, that is, they are all in the open left-half plane, the system is asymptotically stable. The impulse response can be found from the command `impulse(G)`, and will decay to zero as shown in Figure 2.7. The details are given in Script 2.8.

MATLAB Script

```
% Script 2.8:  find poles and impulse response
numG = [1.5 1]                  % generate G(s) as TF object
denG = [1 2 2.5 0.5]            % ...with numerator & denominator
pGroots = roots(denG)           % poles as roots of denominator
G = tf(numG,denG)               % TF object defined
pG = pole(G)                    % poles from 'pole' command
pzmap(G)                        % shows all poles in LHP
impulse(G)                      % impulse response shows stability
```

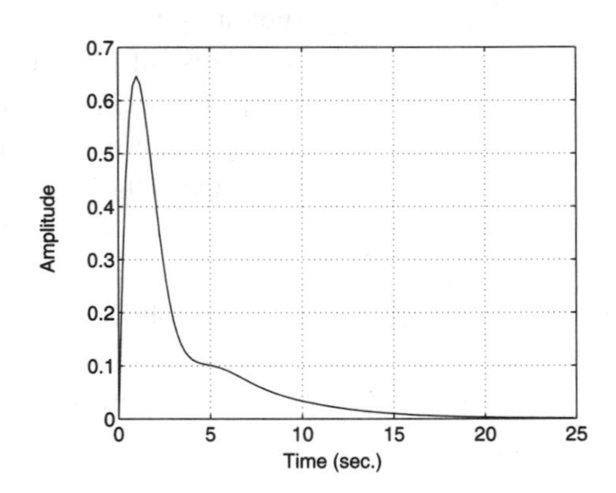

FIGURE 2.7 *Impulse response for Example 2.8*

CROSS-CHECK Verify that the poles of $G(s)$ can be found by using the function `tfdata` to extract the numerator and denominator polynomials from the TF object `G`. ■

WHAT IF?

a. Repeat Example 2.8 if the constant term in the denominator of $G(s)$ is changed from 0.5 to 0. This system now has a pole at the origin, which is not part of the open left half-plane.

b. Repeat Example 2.8 if the constant term in the denominator of $G(s)$ is changed from 0.5 to -0.5. This system now has a pole in the open right-half plane, that is, the right-half plane excluding the imaginary axis. ■

REINFORCEMENT PROBLEMS

In each of the following problems, find the poles of the given transfer function $G(s)$ and determine the stability of the system. Verify the stability by simulating the impulse response. If a system is unstable, provide a time vector from $t = 0$ to 10 seconds as the third input to the `impulse` function so the scaling of the plot will allow the initial transients to be distinguished.

P2.15 Fourth-order system.

$$G(s) = \frac{s^2 + 3s + 4}{s^4 + 8s^3 + 30s^2 + 76s + 80}$$

P2.16 Poles on imaginary axis.

$$G(s) = \frac{s^3 + s^2 + s + 1}{s^4 + 6s^3 + 17s^2 + 54s + 72}$$

P2.17 Repeated poles on imaginary axis.

$$G(s) = \frac{s^2 + 3s + 1}{s^4 + 18s^2 + 81}$$

P2.18 Poles in right half-plane.

$$G(s) = \frac{s^2 + 3s + 4}{s^4 + 4s^3 + 6s^2 + 44s + 80}$$

EFFECTS OF ZEROS ON SYSTEM RESPONSE

The zeros z_j of $G(s)$ do not affect the system stability. However, they do affect the amplitudes of the mode functions in the system response and can block the transmission of certain input signals. In general, to assess the impact of the zeros on the amplitude of the mode functions, a partial-fraction expansion is performed on the Laplace transform of the output signal, and the residue for each mode function is computed. However, certain properties of zeros can be readily illustrated in the time domain as shown in Examples 2.9 and 2.10.

EXAMPLE 2.9
Blocking Zeros

Build a ZPK object for the system whose transfer function is

$$G(s) = \frac{(s + 5)(s^2 + 1)}{(s + 1)(s + 2)^2(s + 3)}$$

Then extract the numerator and denominator polynomials and use the `roots` command to calculate its zeros, and verify that they agree with $G(s)$. Finally, find the time response of the system's output $y(t)$ to the input $u(t) = \sin t$ for $t \geq 0$. Does $y(t)$ contain any sinusoidal component? Explain.

Solution

In MATLAB Script 2.9, we define `zz = [-5; i; -i]`, and `pp = [-1; -2; -2; -3]` and use them to create `G` as a ZPK object.[2] Note that the gain is set to unity by making the third input argument of the `zpk` command be `1`. After drawing the pole-zero plot, we use the `tfdata` command to extract the numerator and denominator polynomials from `G`, yielding `numG = [0 1 5 1 5]` and `denG = [1 8 23 28 12]`. Finally, we use the `roots` command on the numerator polynomial `numG` to calculate the zeros of $G(s)$, which agrees with the original values entered as `zz`.

To obtain the time response, we first generate the time vector `t = [0:0.1:10]'` and the input vector `u = sin(t)`. Then the command `lsim(G,u,t)` is used to simulate the time response $y(t)$, which is shown in Figure 2.8. Note that $y(t)$ consists solely of the modes of $G(s)$ and does not exhibit any sinusoidal component due to the input $u(t)$. To understand this phenomenon, we obtain the Laplace transform of $u(t)$ as

$$U(s) = \mathcal{L}\{u(t)\} = \mathcal{L}\{\sin(t)\} = \frac{1}{s^2 + 1}$$

Then the Laplace transform of the output $y(t)$ is the product

$$Y(s) = G(s)U(s) = \frac{s + 5}{(s + 1)(s + 2)^2(s + 3)}$$

Note that the zeros of $G(s)$ at $s = \pm j1$ cancel the poles of $U(s)$. The response $y(t)$ can be found from the command `impulse(G1,t)` and will be identical to the response obtained from the `lsim` command. The MATLAB commands in Script 2.9 will compute the time response and make the plots.

---

[2]MATLAB will accept either `i` or `j` as $\sqrt{-1}$. Although we use j in the text, we use `i` in the scripts and code fragments because MATLAB uses `i` when it displays complex numbers.

MATLAB Script

```
% Script 2.9: blocking zeros
zz = [-5; i; -i]                % generate G(s) in ZPK form
pp = [-1; -2; -2; -3]
G = zpk(zz,pp,1)                % build G(s) as ZPK object
pzmap(G);grid                   % plot poles & zeros in s-plane
axis([-5 1 -2 2])               % get complex zeros off plot boundary
[numG,denG] = tfdata(G,'v')     % zeros determined as roots
zGroots = roots(numG)           % ....of numerator polynomial
zG = tzero(G)                   % zeros determined by 'tzero' command
t = [0:0.1:10]';                % generate time array
u = sin(t);                     % generate input signal
lsim(G,u,t)                     % simulate system response
[numG,denG] = tfdata(G)         % extract numerator polynomial of G
G1 = tf([1 5],denG)             % remove complex zeros from numG
impulse(G1,t)                   % display impulse response
```


FIGURE 2.8 *Time response for Example 2.9*

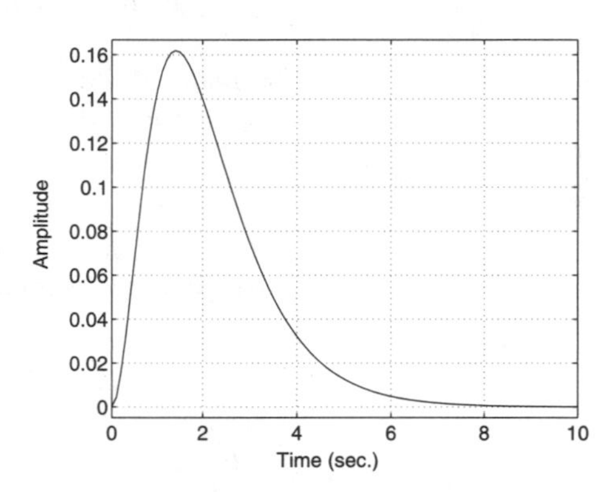

WHAT IF? Suppose the frequency of the sinusoidal input is changed to $\omega = 0.1$, 1.01, and 2 rad/s. Simulate the responses of the system in Example 2.9 and observe that there will be sinusoidal components in $y(t)$. Also note that the sinusoidal component in $y(t)$ will be the smallest for $\omega = 1.01$, since the system attenuates signals with frequencies close to $\omega = 1$. ■

REINFORCEMENT PROBLEMS

In each of the following problems, find the output $y(t)$ for the transfer function $G(s)$ given the input signal $u(t)$. First define the time vector t and the input signal $u(t)$, and use `lsim` to determine $y(t)$. Then perform appropriate zero-pole cancellations, and use `impulse` to determine $y(t)$.

P2.19 Exponentially decaying sinusoidal input.

$$G(s) = \frac{(s+3)(s^2+2s+5)}{(s+1)(s+2)(s+4)^2} \quad \text{and} \quad u(t) = \epsilon^{-t}\sin 2t \text{ for } t > 0$$

P2.20 Derivative function.

$$G(s) = \frac{s}{s+1} \quad \text{and} \quad u(t) = \text{unit-step function}$$

Note that for small s, $G(s) \approx s$, which is the Laplace transform of the derivative operation. What is the steady-state value of the output? Because the input is the unit-step function, the command `step` can be used instead of `lsim`.

CHAPTER 2

EXAMPLE 2.10 ***Right Half-Plane Zero***

Find the zero and the poles of the hydro-turbine system described in Appendix A. Let the incremental gate position ΔQ be a unit-step function and simulate the response of the incremental output power ΔP. Plot the poles and zero of the system in the s-plane.

Solution

The data of the hydro-turbine system are in the file `hydro.m` and can be loaded into the MATLAB workspace by entering the command `hydro`. Alternatively, we can directly create the model as the TF object `G`, as shown in Script 2.10. Extracting `[zz,pp,kk]` from `G`, we find that the zero is $s = 0.5$, which is in the right half-plane, and the poles are $s = -1$ and -2, which are stable. Systems with zeros in the right half-plane are known as *non-minimum phase* systems. The step response is obtained from `step(G)` and is shown in Figure 2.9. Note that the incremental output power Δp initially decreases before it eventually reaches a steady-state value which is positive. This behavior is characteristic of a non-minimum phase system with a single right half-plane zero.

MATLAB Script

```
% Script 2.10: right half-plane zero, hydro-turbine system
G = tf([-2 1],[0.5 1.5 1])          % build model as TF object
[zz,pp,kk] = zpkdata(G,'v')         % system has zero at s = +0.5
step(G)                         % step response shows effect of RHP zero
```

WHAT IF? Change `numG` in Example 2.10 to `numG = [-0.2 1]` such that the zero moves to $s = 5$ and repeat the step-response calculation. In the resulting plot, note the smaller initial decrease in the incremental output power ΔP. This illustrates that moving the right half-plane zero away from the origin reduces its effect on the system response. ■

FIGURE 2.9 *Time response for Example 2.10*

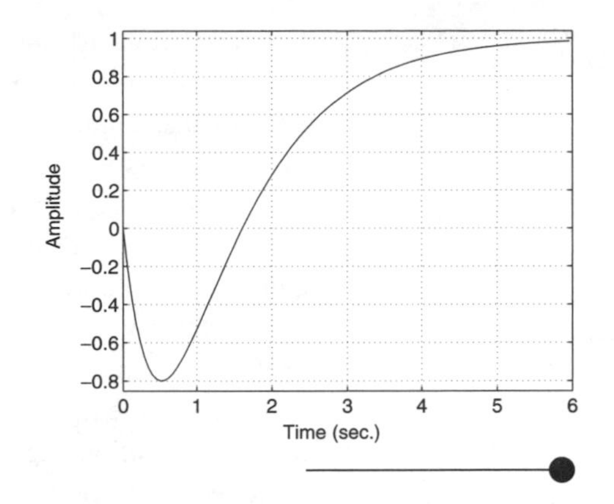

COMPREHENSIVE PROBLEMS

EXPLORATION

E2.1 Interactive Exploration. The file `one_blk.m` is a menu-driven program written in MATLAB that assists the user in building a one-block system model as an LTI object in any of several forms, TF, ZPK, or SS (the state-space form is discussed in Chapter 4). Once the LTI object has been created, the user can select various types of analysis from a menu, including the plotting of step and impulse responses and poles and zeros in the s-plane. It is also possible to obtain the time response to arbitrary inputs. The default system model is that from Example 2.1. However, the user can enter new values at the prompt and the most recently used values are remembered, so they can be reused merely by hitting the Enter key or they can be modified easily.

You are encouraged to use this program by entering the command `one_blk` and letting your curiosity direct you where to go with it. Some suggestions for topics that have been investigated in the examples of this chapter are pole-zero cancellations, the effect of a right half-plane zero on the step response, the relationship of the imaginary part of a complex pole on the frequency of oscillation, and the effect of the real part of a pole on the rate of decay (or growth) of the envelope of the response.

COMPREHENSIVE PROBLEMS

The following problems differ from the others in two ways: (i) they all relate to the "real-world" system models that are described in Appendix A, and (ii) you are asked to apply the MATLAB tools that have been introduced in the chapter. Specifically, for the designated system model, you should run the appropriate M-file and make selections from the menu(s) that appear to generate the requested model as a TF object by starting with a pair of row vectors that contain the coefficients of the transfer function. These vectors of coefficients will always be named `numG` and `denG`, for the numerator and denominator polynomials, respectively. After the TF object has been computed, MATLAB will execute a `keyboard` command so the keyboard will be active and you will be able to enter any of the MATLAB commands.

Because you will be starting with a model in TF form in each case, you can use the `zpkdata` command, with second argument set to `'v'`, to see the zeros and poles of the transfer function $G(s)$ or use the `residue` command to obtain its partial-fraction expansion. You can use the `step` and `impulse` commands to generate and plot the step and impulse responses. Keep in mind that if you use these commands without specifying a time vector, MATLAB will select one. Depending on the particular system with which you are dealing, the time interval that MATLAB uses may or may not make sense. So, be prepared to define your own time vector and use it as an optional argument.

In the problem statements, we will identify the system model to be considered and specify one or more of its transfer functions to be used for analysis. You should feel free to use any of the other transfer functions that we have not specifically mentioned.

CP2.1 Electric power generation system. Run the MATLAB file `epow.m` to obtain the transfer functions from the one input (u) to the three outputs (V_{term}, ω, and P). Verify that the poles of all three transfer functions are $s = -0.105, -0.479 \pm j9.335, -3.081, -26.76, -35.40$, and -114.7. Show that the zeros of each of the transfer functions are different from one another. Do a partial-fraction expansion and look at the relative weightings of the different mode functions.

Examine the response of each of the outputs to a unit step function in the field voltage and relate the character of the response to the dominant poles, which are $s = -0.105$ and the complex pair at $s = -0.479 \pm j9.335$. Note that the real pole should result in a time constant of $\tau = 1/0.105 \approx 10$ s and the pair of complex poles should result in lightly damped oscillations having a period of $T_p = 2\pi/9.335 = 0.673$ s. Replot the oscillatory responses with a time scale such that you can measure the period of the oscillation. Use MATLAB to compute the DC gain for each of the transfer functions and use the results to check the steady-state responses.

CP2.2 Satellite. The MATLAB file `sat.m` builds the model of the satellite and reaction wheel in terms of four transfer functions in TF form. The inputs are the torque applied to the reaction wheel by an electric motor (u) and an external torque (w) that can represent either a disturbance torque applied to the satellite or the torque caused by the firing of the satellite's gas jets when angular momentum is being dumped. Find the zeros, poles, and gain of the four transfer functions and do partial-fraction expansions; compare the results for consistency.

The file `sat.m` will produce plots for the responses to a step in the motor torque and an impulsive external torque. In the first case, the total angular momentum of the satellite and the reaction wheel must remain constant because only internal torques are acting. Use the TF form of the transfer function from u to the wheel speed (Ω) to check the steady-state value of the step response. Note that the initial value of the wheel speed following the impulsive external torque is 7.0 rpm. Can you calculate this value independently by using the values of the systems parameters and a knowledge of what is happening to the angular momentum?

CP2.3 Stick balancer with rigid stick. Run the file `rigid.m` to develop the four transfer functions of the stick balancing system with a rigid stick. The two inputs are the voltage applied to the motor that drives the cart (u) and the disturbance force applied to the stick (w). The two outputs for which transfer functions are developed in `rigid.m` are the angle of the stick (θ) and the position of the cart (x). Two other outputs for which transfer functions could be computed are the angular velocity of the stick (ω) and the velocity of the cart (v). Examine the poles and zeros of each transfer function and comment on their similarities and differences. Also do a partial-fraction expansion and relate its results to the poles and zeros. Try to lend some physical significance to the poles. For example, why is there a pole in the right half-plane, and how does its value relate to the time responses?

Examine the time responses that the file `rigid.m` produces when the inputs are (i) a step in the voltage applied to the motor that drives the cart, and (ii) an impulsive horizontal force applied to the stick. Using the sign conventions shown in Figure A.6 (see Appendix A), give physical arguments supporting that the directions of the translational and rotational motions are correct. Use the `lsim` command to compute and plot the zero-input responses to a small initial angle of the stick or a small initial velocity of the cart.

CP2.4 Stick balancer with flexible stick. Repeat the process described in Problem CP2.3 for the model of the system with a flexible stick by running the file `flex.m`. Compare the poles of the system with a rigid stick with those for the flexible stick. Which poles are the same and which are different? Observe the oscillations in the response of the measured stick angle to the step in motor voltage, and relate the period of the oscillations to the imaginary part of the complex poles.

CP2.5 Hydro-turbine system. In Example 2.10 we observe that the step response of this second-order non-minimum phase system initially goes in the negative direction before becoming positive. To see mathematically why this happens, do a partial-fraction expansion and observe that the residue at one of the two poles is negative. Use the RPI function `rpole2t` to compute and plot the parts of the step response due to each of the poles.

SUMMARY

We have used MATLAB to build transfer-function models of single-block systems as TF and ZPK objects and to convert from one form to the other. We have shown how the system's zeros, poles, and gain can be extracted from the TF form and how the numerator and denominator polynomials can be extracted from the ZPK form.

We have also shown how MATLAB and its Control System Toolbox can be used to perform a variety of analytical operations on these LTI objects, regardless of the form in which they were created. Specifically, we have computed and plotted the system's responses to step functions, impulses, and general inputs, and responses due to individual poles. Also, we have shown how to obtain the contributions to the total response of individual poles, both real and complex, and have looked at the effects of having repeated poles. Finally, we have demonstrated the effects of the transfer-function's poles on the stability of the system and the effect of its zeros on the response.

With these fundamentals established, we are now prepared to consider the construction and analysis of system models that involve more than one block. These topics will be the subject of Chapter 3.

MATLAB FUNCTIONS USED

Function	*Purpose and Use*	*Toolbox*
axis	**axis([xmin xmax ymin ymax])** specifies the plotting area. **axis equal** forces uniform scaling for the real and imaginary axes.	MATLAB
conv	Given two row vectors containing the coefficients of two polynomials, **conv** returns a row vector containing the coefficients of the product of the two polynomials.	MATLAB
cpole2t	Given a complex pair of transfer-function poles, the residues at those poles, and a time vector, **cpole2t** returns the time response due to those poles.	RPI function

CHAPTER 2

dcgain	Given an LTI object of a continuous system, **dcgain** returns the steady-state gain of the system.	Control System
find	**find** returns the indices and values of the nonzero elements of its argument.	MATLAB
get	**get** returns the properties of an LTI object.	MATLAB
impulse	Given an LTI object of a continuous system, **impulse** returns the response to a unit-impulse input.	Control System
lsim	Given an LTI object of a continuous system, a vector of input values, a vector of time points, and possibly a set of initial conditions, **lsim** returns the time response.	Control System
pole	Computes the poles of an LTI object.	Control System
pzmap	Given an LTI object of a continuous system, **pzmap** produces a plot of the system's poles and zeros in the s-plane.	Control System
residue	Given a rational function $T(s) = N(s)/D(s)$, **residue** returns the roots of $D(s) = 0$, the partial-fraction coefficients, and any polynomial term that remains.	MATLAB
roots	Given a row vector containing the coefficients of a polynomial $P(s)$, **roots** returns the solutions of $P(s) = 0$.	MATLAB
rpole2t	Given a real transfer-function pole, the residue at that pole, and a time vector, **rpole2t** returns the time response due to that pole.	RPI function
step	Given a TF model of a continuous system, **step** returns the response to a unit-step function input.	Control System
tf	Given numerator and denominator polynomials **tf** creates the system model as a TF object. The command also converts zero-pole-gain or state-space models to TF form.	Control System
tfdata	Given an LTI object, **tfdata** extracts the numerator and denominator polynomials and other information about the system.	Control System
tzero	Given an LTI object, **tzero** returns the zeros of its transfer function.	Control System

zpk	Given a system's zeros, poles, and gain, **zpk** creates the system model as a ZPK object. The command also converts transfer-function or state-space models to ZPK form	Control System
zpkdata	Given an LTI object, **zpkdata** extracts the zeros, poles, and gain and other information about the system.	Control System

ANSWERS

P2.1 Zeros are $s = -0.5, -1.0$; poles are $s = -2.0, -0.20 \pm j0.9798$; gain $= 2.0$

P2.2 Zeros are $s = -1.0 \pm j0.5774$; poles are $s = -0.2504 \pm j0.4980$, $-1.2496 \pm j2.2082$; gain $= 3.0$

P2.3 $G(s) = (150s^2 + 900s + 1200)/(s^4 + 29s^3 + 240s^2 + 1252s + 1040)$

P2.4 $G(s) = (5s^3 + 95s^2 + 580s + 1020)/(s^4 + 45s^3 + 504s^2 + 1660s + 1200)$

P2.5 Poles are $s = -2.0 \pm j4.0$ and -3.0 and the residue of the pole at $s = -3.0$ is -0.5882. The impulse response has a maximum value of 2.0 at $t = 0+$ and a minimum value of $y = -0.67$ at $t = 0.70$.

P2.6 Poles are $s = -0.50 \pm j1.3229, -0.50$, and -2.0. Residue at $s = -0.50 + j1.3229$ is $0.0179 - j0.1417$. The impulse response peaks at $h = 0.51$ at $t = 0.9$ and then decays to zero as $t \longrightarrow \infty$, but it does not do so monotonically.

P2.9

$$G(s) = \frac{-0.8889}{s + 0.75} + \frac{6.6667}{(s + 1.5)^2} + \frac{0.8889}{s + 1.5}$$

P2.10

$$G(s) = \frac{2.3333}{(s + 0.5)^2} + \frac{2.2222}{s + 0.5} - \frac{2.6667}{(s + 2)^2} - \frac{2.2222}{s + 2}$$

P2.11 Step response reaches a maximum value of 0.375 at $t = 0.36$ s and approaches a steady-state value of $y_s(\infty) = 5/60 = 0.0833$.

P2.12 Step response increases monotonically from 0 to a steady-state value of 1.25.

P2.13 The response reaches a maximum of 0.61 at $t = 2.1$ s and settles to its final value of 0.333 by $t = 5$ s.

P2.14 The response reaches a maximum of 0.59 at $t = 1.25$ s, a minimum of -0.30 at $t = 2.8$ s, and settles to its final value of 0 just after $t = 5$ s.

P2.15 Poles are $s = -1.0 \pm j3.0, -2.0$, and -4.0. The system is stable and the plot shows that $h(t) \approx 0$ for $t > 5$ s.

P2.16 Poles are $s = \pm j3.0, -2.0$, and -4.0. The system is marginally stable and the plot of $h(t)$ shows an undamped oscillation with an amplitude of 0.45, a period of $2\pi/3.0 = 2.094$ s, and an average value of 0.

P2.17 Poles are $s = \pm j3.0$, with a multiplicity of 2. The plot of $h(t)$ is a sinusoid of period of $2\pi/3.0 = 2.094$ s whose amplitude grows *linearly* with time.

P2.18 Poles are $s = +1.0 \pm j3.0, -2.0$, and -4.0. The plot of $h(t)$ contains a sinusoid that grows *exponentially* with time.

P2.19 The response rises rapidly from 0 to a maximum of 0.091 at $t = 0.9$ s and then decays monotonically, becoming very close to its final value of 0 by $t = 8$ s.

P2.20 The response starts at 1.0 and decays monotonically to its final value of 0, with a time constant of 1.0 s.

CHAPTER 3

Building and Analyzing Multi-block Models

PREVIEW

A control system is usually described in terms of an interconnection of components. Each component may be described by a set of differential equations or a transfer function. Based on the interconnection information, often given in the form of a block diagram, a designer has to build the control system model from the descriptions of the component models. We discuss in this chapter three basic model interconnections—series, parallel, and feedback—and show how to interconnect models given in either transfer-function or zero-pole-gain form using commands from the Control System Toolbox. We also illustrate the property that some interconnections preserve the zeros of the component models, whereas others preserve the poles of the component models. Interconnections for models given in state-space form will be discussed in Chapter 4.

SERIES CONNECTIONS

Figure 3.1 shows two systems that are connected in series (also referred to as a cascade connection). Provided that no loading effects are present, the transfer function of the resulting system is the product of the individual transfer functions, namely $T(s) = G_2(s)G_1(s)$. Note that we have placed the transfer function of the left-hand block *after* that of the right-hand block. The reason

FIGURE 3.1 *Two systems connected in series*

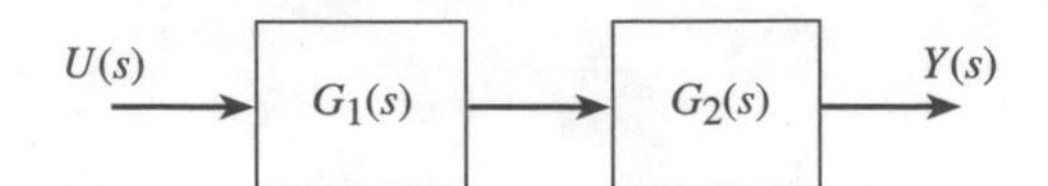

for doing this is evident if we define the output of $G_1(s)$ to be $V(s) = G_1(s)U(s)$ and express the output of the second block as $Y(s) = G_2(s)V(s)$. Using the first equation to substitute for $V(s)$ in the second equation, we see that the proper expression for the transform of the output is

$$Y(s) = G_2(s)G_1(s)U(s)$$

which says that the transfer function of the series connection is

$$T(s) = G_2(s)G_1(s) \tag{3.1}$$

For single-input/single-output (SISO) systems in either TF or ZPK form, or a mixture of the two forms, the results will be the same for $G_2(s)G_1(s)$ or $G_1(s)G_2(s)$. However, if the state-space form is used, as will be done in Chapter 4, the internal representations, in terms of the state-space matrices, will be different. More important, for systems that are not SISO, the operation $G_1(s)G_2(s)$ may not even exist, and will most likely give incorrect results if it does. Thus, we will use the expression in (3.1) for the series connection depicted in Figure 3.1.

It follows that the poles of $T(s)$ are the combined poles of $G_1(s)$ and $G_2(s)$ and that the zeros of $T(s)$ are the combined zeros of $G_1(s)$ and $G_2(s)$. Should a zero of $G_2(s)$ coincide with a pole of $G_1(s)$ or a zero of $G_1(s)$ be the same as a pole of $G_2(s)$, then the pole-zero pair would cancel and neither would show up in $T(s)$. Should both $G_1(s)$ and $G_2(s)$ have an identical pole or zero, then the multiplicity of that pole or zero would be increased accordingly.

Since the poles of $T(s)$ are the combined poles of $G_1(s)$ and $G_2(s)$, the series connection $T(s)$ has the same stability property as $G_1(s)$ and $G_2(s)$, provided that they do not have common poles on the $j\omega$-axis. The stability property of $T(s)$ should be determined before any pole-zero cancellation. For example, if $G_1(s)$ has a pole at $s = 1$ and $G_2(s)$ has a zero at $s = 1$, then $T(s)$ is (internally) unstable even though the instability is not observed at the output of the interconnection.

Expressing linear time-invariant systems as LTI objects has the advantage that system interconnections become more transparent. For example, the series connection of two systems $G_1(s)$ and $G_2(s)$ shown in Figure 3.1 is mathematically expressed as $G(s) = G_2(s) \times G_1(s)$. In MATLAB 5, the series connection is obtained by the command `G = G2*G1`, where `G1` and `G2` are LTI objects in any form. When operating on two LTI objects, the symbol $*$ means *series* connection. Since the interpretation of $*$ is based on the properties of its operands, it is an example of an *overloaded operator*.

EXAMPLE 3.1 *Series Connection of Two Systems*

Create the series connection, denoted by the transfer function $T(s)$, of the two systems whose transfer functions are $G_1(s)$ and $G_2(s)$, where

$$G_1(s) = \frac{2s + 3}{5s^2 + 2s + 2}$$

and $G_2(s)$ has a zero at $s = -2$, poles at $s = -0.5$ and -8, and a gain of 5. Give both the TF and ZPK forms of the series connection and plot the unit-step response. Determine the stability of the series connection.

Solution

The MATLAB commands in Script 3.1 start by building $G_1(s)$ as a TF object. Then $G_2(s)$ is built in ZPK form. Finally, the * operator is used to connect the two systems, resulting in the transfer function $T(s)$. The zeros and poles of $T(s)$ are found by using the command `zpkdata`.

MATLAB Script

```
% Script 3.1:  Series connection of two blocks
G1 = tf([2 3],[5 2 2])                  % G1(s) in TF form
G2 = zpk(-2,[-0.5;-8],5)                % G2(s) in ZPK form
Tzpk = G2*G1                            % series combination in ZPK form
Ttf = tf(Tzpk)                          % series combination in TF form
step(Tzpk), hold on                     % step response of T(s) from ZPK form
step(Ttf), hold off                     % step response of T(s) from TF form
[zT,pT,kT] = zpkdata(Tzpk, 'v')         % zeros, poles, & gain from ZPK form
```

The results are

$$T(s) = \frac{10s^2 + 35s + 30}{5s^4 + 44.5s^3 + 39s^2 + 25s + 8} \qquad (3.2)$$

which has zeros at $s = -1.5$ and -2, poles at $s = -0.2 \pm j0.6, -0.5$, and -8, and a gain of 2. Since the poles of $G_1(s)$ and $G_2(s)$ are all in the open left half-plane, the series connection is stable.

Comment: We can see that the zeros of $T(s)$ are those of $G_1(s)$ [$s = -1.5$] and $G_2(s)$ [$s = -2$]. Likewise, the poles of $T(s)$ are the poles of $G_1(s)$ [$s = -0.2 \pm j0.6$] and the poles of $G_2(s)$ [$s = -0.5$ and -8]. Finally, the gain of $T(s)$ is the product of the gain of $G_1(s)$ and the gain of $G_2(s)$ [$0.4 \times 5 = 2$].

REINFORCEMENT PROBLEMS

Use MATLAB to make series connections of the type shown in Figure 3.1, except where noted to the contrary, of the systems whose transfer functions are given below. In each case create the transfer function $T(s)$ of the combined system as a ratio of polynomials and determine its zeros, poles, and gain. Determine the stability of $T(s)$ and plot the response to the unit step function.

P3.1 Two second-order systems.

$$G_1(s) = \frac{3s^2 + 2s + 1}{4s^2 + 5s + 8} \quad \text{and} \quad G_2(s) = \frac{4s + 2}{s^2 + 2s + 10}$$

P3.2 Third- and first-order systems.

$$G_1(s) = \frac{2(s + 4)}{(s + 1)(s^2 + 4s + 16)} \quad \text{and} \quad G_2(s) = \frac{5}{s + 8}$$

P3.3 Interchanging two systems. Repeat Example 3.1, but with the order of the two systems reversed, so the block $G_2(s)$ precedes the block $G_1(s)$. Verify that the resulting system has the same transfer function as in the example.

P3.4 Three systems in series. $T(s) = G_1(s)G_2(s)G_3(s)$ where

$$G_1(s) = \frac{2s + 1}{s + 5}, \qquad G_2(s) = \frac{5(2s + 3)}{5s^2 + 2s + 2}, \quad \text{and} \quad G_3(s) = \frac{3s + 1}{s^2 + 2s + 8}$$

P3.5 Stable zero-pole cancellation.

$$G_1(s) = \frac{4s + 6}{s^2 + 5s + 10} \quad \text{and} \quad G_2(s) = \frac{2(s + 4)}{s^2 + 7.5s + 9}$$

P3.6 Common zero and pole in right half-plane.

$$G_1(s) = \frac{s^2 + 2.5 + 1.5}{s^3 + 3.1s^2 - 3.9s - 12.6} \quad \text{and} \quad G_2(s) = \frac{s - 2}{s^2 + 2s + 5}$$

PARALLEL CONNECTIONS

Two single-input/single-output systems can be connected in parallel by joining their inputs and summing their outputs, as indicated in Figure 3.2. The transfer function of the resulting system will be

$$T(s) = G_1(s) + G_2(s)$$

which will have poles that are the union of the poles of $G_1(s)$ and $G_2(s)$. However, the zeros of $T(s)$ will differ from the zeros of $G_1(s)$ and $G_2(s)$.

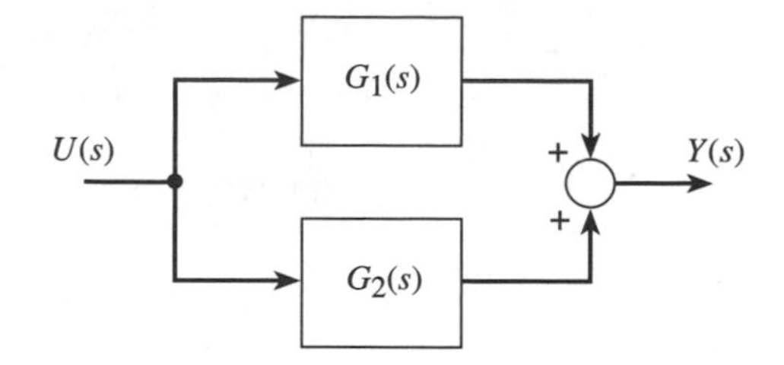

FIGURE 3.2 *Two systems connected in parallel*

Since the poles of $T(s)$ are the union of the poles of $G_1(s)$ and $G_2(s)$, then the parallel connection $T(s)$ has the same stability property as $G_1(s)$ and $G_2(s)$. This stability property is not affected by $G_1(s)$ and $G_2(s)$ having common poles on the $j\omega$-axis, as in the series connection. Should a zero of $T(s)$ be the same as a pole of $G_1(s)$ or $G_2(s)$, then the pole-zero pair cancel. However, system stability should be determined before any pole-zero cancellations.

In MATLAB the parallel connection of two systems is mathematically expressed as $T(s) = G_1(s) + G_2(s)$ and is obtained by the command `T = G1 + G2`. When working with two LTI objects, the symbol + is an overloaded operator meaning *parallel* connection. The following example will illustrate this process.

EXAMPLE 3.2
Parallel Connection

Form the parallel connection of the two transfer functions

$$H_1(s) - \frac{2s + 6}{s^2 + s + 8} \quad \text{and} \quad H_2(s) = \frac{s + 4}{(s + 1)(s^2 + 4s + 1)}$$

Determine the zeros, poles, and gain of $T(s)$ and compare them with the corresponding parameters of $H_1(s)$ and $H_2(s)$. Also determine the stability of $T(s)$ and plot the step response.

Solution

A set of MATLAB commands that will accomplish the objectives of this example appear in Script 3.2. The values of the zeros, poles, and gain of $H_1(s)$, $H_2(s)$, and $T(s)$ are shown in Table 3.1. It is apparent that the poles of $T(s)$ consist of the poles of $H_1(s)$ and $H_2(s)$, but the zeros of $T(s)$ are different from those of $H_1(s)$ and $H_2(s)$, as expected. Since the poles of $H_1(s)$ and $H_2(s)$ are all in the open left half-plane, the parallel connection $T(s)$ is stable.

MATLAB Script

```
% Script 3.2:  Parallel connection of two blocks
H1 = tf([2 6],[1 1 8])                      % H1(s) in TF form
denH2 = conv([1 1],[1 4 1])                 % denominator of H2(s)
H2 = tf([1 4],denH2)                        % H2(s) in TF form
T = H1 + H2                                 % T(s) in TF form
[zT,pT,kT] = zpkdata(T,'v')                 % zeros, poles, and gain of T(s)
pH1 = pole(H1)                              % poles of H1(s)
pH2 = pole(H2)                              % poles of H2(s)
step(T)                                     % step response of T(s)
```

TABLE 3.1 *Zeros and poles of the transfer functions in Example 3.2*

	$H_1(s)$	$H_2(s)$	$T(s)$
zeros	-3	-4	$-0.4093 \pm j1.0508$, $-3.8407 \pm j0.4347$
poles	$-0.5 \pm j2.7839$	$-0.2679, -1, -3.7321$	$-0.5 \pm j2.7839$, $-0.2679, -1, -3.7321$

CHAPTER 3

WHAT IF? Suppose that in Example 3.2 we had formed the difference of the two transfer functions. Specifically, use the overloaded operator $-$ to create the parallel connection $D(s) = H_2(s) - H_1(s)$ and determine its zeros and poles and plot its response to the unit step function. You should find that although $D(s)$ has the same poles as $T(s)$ in the example, it has a zero in the right half-plane. Its step response starts out in the negative direction, before becoming positive and approaching a positive steady-state value. How would you expect the step response of the system defined by $G(s) = H_1(s) - H_2(s)$ to look? Use MATLAB to check it out. ■

REINFORCEMENT PROBLEMS

Use MATLAB to make a parallel connection of the two systems whose transfer functions are given below. In each case create the transfer function of the combined system as a ratio of polynomials, determine its zeros, poles, and gain, and plot the response to the unit step function. Also determine the stability of the interconnected system.

P3.7 Second- and first-order systems.

$$G_1(s) = \frac{2s + 3}{5s^2 + 2s + 2} \quad \text{and} \quad G_2(s) = \frac{3(s + 5)}{6s + 1}$$

P3.8 Two second-order systems.

$$G_1(s) = \frac{5s + 4}{2s^2 + 4s + 12} \quad \text{and} \quad G_2(s) = \frac{3s^2 + s + 4}{5s^2 + 12s + 3}$$

P3.9 Stable zero-pole cancellation.

$$G_1(s) = \frac{1}{s^2 + 3s + 2} \quad \text{and} \quad G_2(s) = \frac{2s + 5}{s^2 + 5s + 6}$$

Write the transfer function $T(s) = G_1(s) + G_2(s)$ as a ratio of polynomials both before and after any zero-pole cancellations. Do you observe a stable pole-zero cancellation?

P3.10 Common zero and pole in right half-plane.

$$G_1(s) = \frac{-3}{s^2 - s - 2} \quad \text{and} \quad G_2(s) = \frac{2s + 1}{s^2 + s - 6}$$

Write the transfer function $T(s) = G_1(s) + G_2(s)$ as a ratio of polynomials both before and after any zero-pole cancellations. Verify that the instability is not observed when looking at the output of $T(s)$.

EXAMPLE 3.3
Series/Parallel Connections

Use the $*$ and $+$ operators to connect four subsystems in the series/parallel arrangement shown in Figure 3.3 where $G_1(s)$ and $G_2(s)$ in the figure are the two blocks from Example 3.1, and $G_3(s)$ and $G_4(s)$ in the figure are the two blocks from Example 3.2, with $G_3(s) = H_1(s)$ and $G_4(s) = H_2(s)$. Verify that the poles of the overall transfer function are the union of the poles of the individual transfer functions. Also verify that the zeros of the overall transfer function do not include the zeros of $G_1(s), G_2(s)$, or $G_3(s)$ but do include the zero of $G_1(s)$. Show that your step response agrees with that of Figure 3.4.

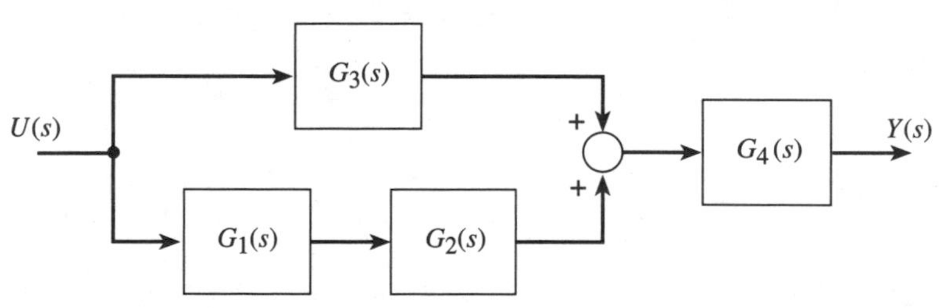

FIGURE 3.3 *A series/parallel connection of four systems*

BUILDING AND ANALYZING MULTI-BLOCK MODELS

FIGURE 3.4 *Step response of series/parallel connection in Example 3.3*

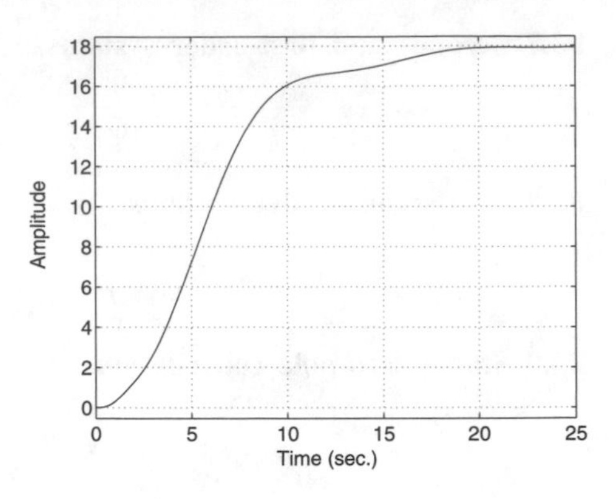

Solution

First, we use the definitions of the transfer functions of the individual blocks to create each of them as LTI objects, with G1, G3, and G4 being in TF form and G2 being in ZPK form. We use MATLAB's convolution function conv to form the denominator polynomial of G4 without creating a separate system or variable. Once this has been done, the series/parallel connection $T(s)$ can be obtained with the single command T = G4*(G3 + G2*G1), resulting in

$$T(s) = \frac{2(s+4)(s+2.833)(s+9.057)(s+0.9049)(s^2+0.1046s+1.24)}{(s+3.732)(s+8)(s+1)(s+0.5)(s+0.2679)(s^2+0.4s+0.4)(s+s+8)}$$

To retrieve the six zeros and nine poles of the overall system as column vectors, we use [zT,pT,kT] = zpkdata(T,'v'). The values are listed in Table 3.3. Then the step command is used to compute and plot its step response shown in Figure 3.4. The dcgain command computes the steady-state value of the step response as 18.0, which agrees with the figure.

The MATLAB commands that will create $T(s)$ and carry out the required analysis are given in Script 3.3.

MATLAB Script

```
% Script 3.3  Series/Parallel connection of four blocks
G1 = tf([2 3],[5 2 2])                        % G1(s) in TF form
G2 = zpk(-2,[-0.5; -8],5)                     % G2(s) in ZPK form
G3 = tf([2 6],[1 1 8])                        % G3(s) in TF form
G4 = tf([1 4],conv([1 1],[1 4 1])             % G4(s) in TF form
T = G4*(G3 + G2*G1)         % series/parallel connection of G1, G2, G3 & G4
[zT,pT,kT] = zpkdata(T,'v')                   % zeros, poles, & gain of T(s)
TgainDC = dcgain(T)                           % DC gain of T(s)
step(T)                                       % step response of T(s)
```

By using the zpkdata command on the LTI object T, we obtain the entries in the following table of zeros and poles.

TABLE 3.2 *Zeros and poles of the transfer functions in Example 3.3*

	$G_1(s)$	$G_2(s)$	$G_3(s)$	$G_4(s)$	$T(s)$
zeros	-1.50	-2.0	-3.0	-4.0	$-0.0523 \pm j1.1124$, -0.9049, -2.8333, -4.0, -9.0572
poles	$-0.20 \pm j0.60$	-0.50, -8.0	$-0.50 \pm j2.7839$	-0.2679, -1.0, -3.7321	$-0.20 \pm j0.60$, -0.2679, -0.50, $-0.50 \pm j2.7839$, -1, -3.7321, -8.0

REINFORCEMENT PROBLEMS

Use MATLAB to make a series/parallel connection according to the block diagram given in Figure 3.3 of the four systems whose transfer functions are given below. In each case create the transfer function of the combined system in TF form, determine its zeros, poles, and gain, and plot the response to the unit step function. Also determine the stability of the interconnected system.

P3.11 Series/parallel connection.

$$G_1(s) = \frac{12s^2 + 5s + 1}{24s^2 + 15s + 6}, \quad G_2(s) = \frac{4(s + 6)}{(s + 2)(s + 20)},$$

$$G_3(s) = \frac{10}{s + 10}, \quad \text{and} \quad G_4(s) = \frac{3s + 2}{4s^2 + 3s + 6}$$

P3.12 Another series/parallel connection.

$$G_1(s) = \frac{6}{s^2 + 5s + 1}, \quad G_2(s) = \frac{4s + 1}{16s + 1},$$

$$G_3(s) = \frac{7s + 0.5}{10s + 3}, \quad \text{and} \quad G_4(s) = \frac{4}{s^2 + 3s + 1}$$

FEEDBACK CONNECTIONS

Figure 3.5 shows two systems connected in a feedback configuration with a negative sign associated with the feedback signal where it enters the summing junction. We refer to $G(s)$ as the forward transfer function and to $H(s)$ as the feedback transfer function. The transfer function of the complete system, denoted by $T(s)$, with the negative feedback as shown in the figure, is

$$T(s) = \frac{G(s)}{1 + G(s)H(s)}$$

The feedback system $T(s)$ can be implemented using the Control System Toolbox command `feedback` which accepts LTI objects as input arguments. For example, the connection of $G(s)$ and $H(s)$ as shown in Figure 3.5 is obtained with the command `T = feedback(G,H)`, where negative feedback is assumed. If the feedback signal enters the summing junction with a + sign, a third argument of +1 is included. For unity feedback, the command would be `T = feedback(G,1)`, in which the second argument 1 represents a system having unit gain.

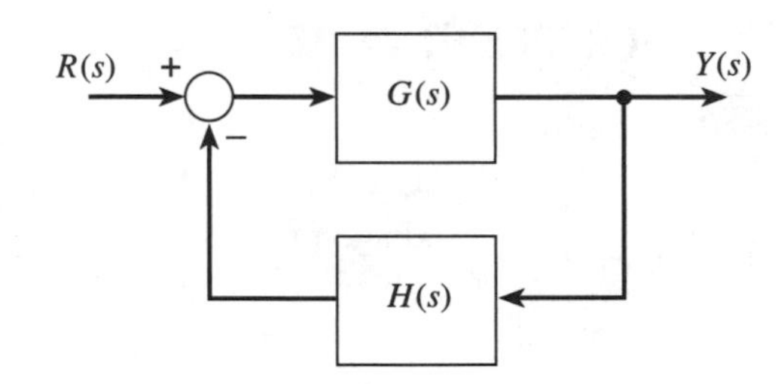

FIGURE 3.5 *A feedback connection of two systems*

It can be shown that the closed-loop transfer function $T(s) = Y(s)/U(s)$ will have zeros that are the zeros of $G(s)$ and the poles of $H(s)$. The closed-loop poles will differ from the poles of $G(s)$ and $H(s)$.

EXAMPLE 3.4 ***Feedback Connection***

Use MATLAB to create the negative feedback system shown in Figure 3.5 where

$$G(s) = \frac{4s + 1}{5s^2 + 3s + 2} \quad \text{and} \quad H(s) = \frac{s + 6}{s^2 + 4s + 11}$$

Give the closed-loop transfer function as a ratio of polynomials and determine its zeros, poles, and gain. Compare the poles of the closed-loop transfer function $T(s)$ to the poles of $G(s)$ and $H(s)$. Determine the stability of $T(s)$. Also verify that the zeros of $T(s)$ are the zeros of $G(s)$ and the poles of $H(s)$.

Solution The MATLAB commands to solve this example are contained in Script 3.4.

MATLAB Script

```
% Script 3.4:  Feedback connection of two blocks
G = tf([4 1],[5 3 2])                % G(s) in TF form
pG = pole(G)                         % poles of G(s)
H = tf([1 6],[1 4 11])               % H(s) in TF form
pH = pole(H)                         % poles of H(s)
T = feedback(G,H)                    % feedback connection in TF form
[zT,pT,kT] = zpkdata(T,'v')          % zeros, poles, & gain of T(s)
step(T)                              % step response of closed-loop system
```

The `feedback` command returns the closed-loop transfer function as

$$T(s) = \frac{4s^3 + 17s^2 + 48s + 11}{5s^4 + 23s^3 + 73s^2 + 66s + 28}$$

Table 3.3 shows the poles and zeros of $G(s)$, $H(s)$, and $T(s)$. It is apparent that the zeros of $T(s)$ are the zeros of $G(s)$ and the poles of $H(s)$, and that the poles of $T(s)$ differ from any of the poles of $G(s)$ and $H(s)$. Since the poles of $T(s)$ are all in the open left half-plane (the left half-plane exclusive of the imaginary axis), the feedback connection $T(s)$ is asymptotically stable.

TABLE 3.3 *Zeros and poles of the transfer functions in Example 3.4*

	$G(s)$	$H(s)$	$T(s)$
zeros	-0.25	-6	$-0.25,\ -2 \pm j2.6458$
poles	$-0.3 \pm j0.5568$	$2 \pm j2.6458,$	$-0.5539 \pm j0.4932,$ $-1.7461 \pm j2.6707$

WHAT IF? Suppose the forward-path transfer function of the feedback system considered in Example 3.4 contains a gain K such that $G(s) = K(4s + 1)/(5s^2 + 3s + 2)$. Calculate the closed-loop zeros and poles and plot the unit step response for several values of the gain K. Observe that K will affect the closed-loop poles but not the closed-loop zeros. Why is that the case? You should find that the closed-loop system will remain stable for values of gain up to about $K = 37$. Verify that the system is stable for a small range of *negative* gains that includes the interval $-1.4 < K < 0$. ■

BUILDING AND ANALYZING MULTI-BLOCK MODELS

REINFORCEMENT PROBLEMS

In Problems 3.13 through 3.19 use MATLAB to determine $T(s)$, the transfer function of the closed-loop system obtained by connecting the systems described by $G(s)$ and $H(s)$ with negative feedback, as shown in Figure 3.5. Write $T(s)$ as a ratio of polynomials and make a table showing the zeros and poles of $G(s), H(s)$, and $T(s)$. Determine the stability of the closed-loop system.

P3.13 Feedback connection.

$$G(s) = \frac{2s + 1}{4s^2 + 5s + 8} \quad \text{and} \quad H(s) = \frac{4s + 7}{(s + 1)(s + 8)}$$

P3.14 Another feedback connection.

$$G(s) = \frac{3(s + 6)}{(s + 1)(s^2 + 3s + 5)} \quad \text{and} \quad H(s) = \frac{5s + 1}{15s + 1}$$

P3.15 One more feedback connection.

$$G(s) = \frac{5s + 12}{(s + 0.5)(s + 10)} \quad \text{and} \quad H(s) = \frac{6}{s^2 + 7s + 3}$$

P3.16 Unstable open-loop system.

$$G(s) = \frac{1}{(s - 2)(s + 8)} \quad \text{and} \quad H(s) = \frac{100(s + 5)}{s + 20}$$

P3.17 Gain variation.

$$G(s) = \frac{K}{s(s^2 + 8s + 32)}, \quad H(s) = \frac{s + 8}{s + 2}, \quad \text{and} \quad K = 1$$

Repeat the problem for $K = 10$, 100, and 1000. Investigate how the closed-loop system poles vary with the controller gain K. You should observe that one of the poles moves toward the zero at $s = -8$, and the other three poles become very large. This type of gain-variation investigation, known as *root locus,* will be discussed in Chapter 5. *Hint:* You can make a vector of the four values of K and create a `for` loop to do the calculations for each of the gains.

P3.18 Gain variation for a second-order system.

$$G(s) = \frac{K(s + 5)}{(s + 1)(s + 3)}, \quad H(s) = 1, \quad \text{and} \quad K = 1$$

Repeat the problem for $K = 10$, 100, and 1000. Investigate how the closed-loop system poles vary as the controller gain K is increased. Will the feedback system remain stable for all positive values of K? Because $H(s) = 1$, you can use the command `cloop` as an alternative to `feedback`.

P3.19 Positive feedback. Repeat Example 3.4 with positive rather than negative feedback. You should find that $T(s)$ has different poles than before, but they are still in the left half-plane. Also $T(s)$ has the same zeros as before.

CONTROLLER TRANSFER FUNCTIONS

In Chapter 9 we will discuss lead-lag controllers and techniques for designing the controller to meet a set of given performance specifications. At this point we will present the transfer functions of these controllers so they can be used with the transfer-function models of our "real-world" systems (see Appendix A) to build models of feedback control systems that can be analyzed for stability properties and step response. These controllers are known as lag, lead, and lead-lag compensators because of the phase characteristics of their frequency responses. The other type of controller we may want to consider is the proportional controller, which consists of just a gain. The transfer functions of these controllers will be expressed in terms of three parameters, namely, the gain K, the location(s) of the zero(s) in the s-plane, and the parameter α, which is the ratio of the pole to the zero, or vice versa. We will define α so it is always > 1, although some authors define it differently so it can also be less than 1.

The transfer function of the lag controller is

$$G_c(s) = K_{\text{lag}}\left(\frac{s - z_{\text{lag}}}{s - z_{\text{lag}}/\alpha_{\text{lag}}}\right) \tag{3.3}$$

which has a pole on the negative real axis at $s = z_{\text{lag}}/\alpha_{\text{lag}}$ and a zero to the left of the pole at $s = z_{\text{lag}}$.

The transfer function of the lead controller is

$$G_c(s) = K_{\text{lead}}\left(\frac{s - z_{\text{lead}}}{s - \alpha_{\text{lead}} z_{\text{lead}}}\right) \tag{3.4}$$

which has a zero on the negative real axis at $s = z_{\text{lead}}$ and a pole to the left of the zero at $s = \alpha_{\text{lead}} z_{\text{lead}}$.

The lead-lag controller can be thought of as a series combination of the lead and lag controllers, with a single gain. Hence, its transfer function will have two zeros and two poles, all on the negative real axis, with the two zeros between the poles. We will use the same value of α for both the lead and the lag portion, so the controller transfer function can be written as

$$G_c(s) = K_{\text{ldlg}}\left(\frac{s - z_{\text{lag}}}{s - z_{\text{lag}}/\alpha}\right)\left(\frac{s - z_{\text{lead}}}{s - \alpha z_{\text{lead}}}\right) \tag{3.5}$$

FEEDBACK SYSTEMS WITH TWO INPUTS

Figure 3.6(a) shows a block diagram of a feedback system having two inputs: (i) the reference input labeled as $R(s)$ and (ii) the disturbance input $D(s)$. Ideally, we would like the output to follow the reference input and to not be affected by the disturbance input. Although it is possible to construct a 2-input/1-output model of such a system by using the `connect` command of the Control System Toolbox, we will instead construct two 1-input models and use the appropriate one depending on which input is under consideration.

When we are interested in the response to the reference input, we can set the disturbance input to zero and obtain the single-input feedback system shown in part (b) of Figure 3.6. In this case the forward transfer function is the product of the controller and plant transfer functions $G_c(s)G_p(s)$, and the feedback transfer function is $H(s)$. With these associations we can use the results of the previous section.

Likewise, when we are interested in the response to the disturbance input, we set $R(s) = 0$ and obtain the single-input block diagram shown in part (c) of Figure 3.6. Now the forward transfer function is $G_p(s)$ and the feedback transfer function is $-G_c(s)H(s)$, and there is a plus sign at the arrow where the feedback signal enters the summing junction. An equivalent diagram that retains the feature of negative feedback can be had by associating the negative sign in the feedback path with the summing junction, as seen in part (d) of Figure 3.6.

We will illustrate the construction of a model of a feedback system having two inputs in the following example. The poles of the closed-loop system are unaffected by our choice of input, but the closed-loop zeros are affected. Put another way, the mode functions of the closed-loop system are the same regardless of the choice of input, but the weightings of these mode functions in the response are different.

CHAPTER 3

EXAMPLE 3.5
2-Input Feedback System

Consider a feedback system that has both reference and disturbance inputs, as shown in Figure 3.6(a), where the plant and sensor transfer functions are

$$G_p(s) = \frac{4}{(0.5s + 1)(2s + 1)} \quad \text{and} \quad H(s) = \frac{1}{0.05s + 1}$$

and the controller is a lead controller (3.4) with a gain of 80, a zero at $s = -3$, and a pole/zero ratio of $\alpha = 15$. Develop two separate MATLAB models, one model having the input $R(s)$, and the other having the input $D(s)$. Then use these models to determine the closed-loop zeros and poles and to plot their responses to a unit step function on a single set of axes.

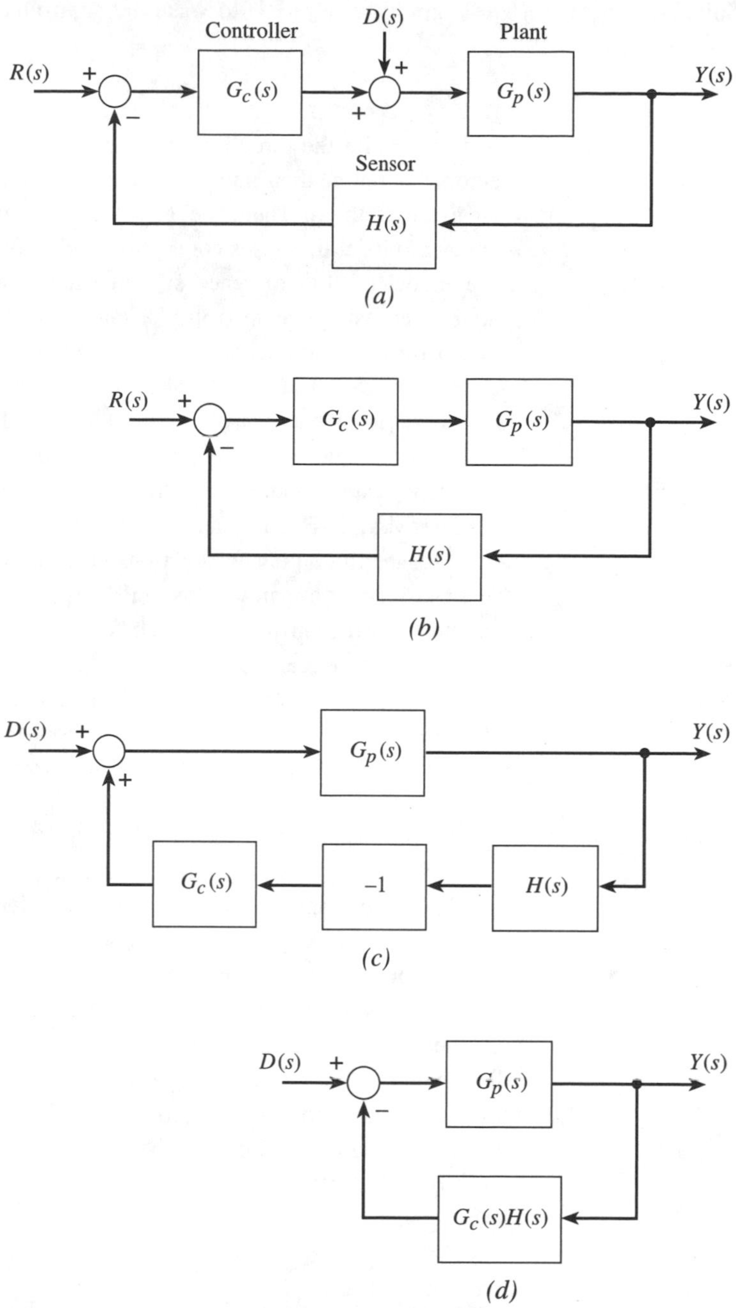

FIGURE 3.6 *Block diagrams of a feedback system with reference and disturbance inputs (a) Complete system (b) Single-input model for reference input only (c) Single-input model for disturbance input only (d) Alternate model for disturbance input only*

Solution We begin by using (3.4) to write the controller transfer function as

$$G_c(s) = \frac{80(s + 3)}{s + 45}$$

Next, we create the three individual blocks in TF objects by entering the row vectors that define their numerator and denominator polynomials or expressions that evaluate them. Then, the `feedback` command can be used with the * operator to form the series connection in the forward path to obtain the closed-loop model of the reference system shown in Figure 3.6(b). Care must be taken when using the `feedback` command to be sure that the correct sign is used for the feedback signal at the summing junction. If a negative sign is desired, it does not have to be shown explicitly. However, if positive feedback is desired, as in part (c) of Figure 3.6, then a third argument must be used that has the value +1. The `feedback` command is also used to create the disturbance-input model, using the alternate form shown in Figure 3.6(d).

The MATLAB commands in Script 3.5 implement the steps described above for the three transfer functions given in the example statement. By using the `zpkdata` command on the ZPK objects of the closed-loop systems, we find the closed-loop poles of both to be $s = -3.4670, -7.3336 \pm j7.9786$, and -49.3658. Recall that the closed-loop zeros are the *zeros* of the *forward* transfer function and the *poles* of the *feedback* transfer function. For the reference transfer function, the closed-loop zeros are $s = -3.0$ and -20, where $s = -3.0$ is the zero of the forward transfer function [because $G_c(s)$ is in the forward path], and $s = -20$ is the pole of the sensor transfer function. For the disturbance transfer function, the zeros are $s = -45$ and -20. The value of $s = -45$ results from the fact that $G_c(s)$ is in the feedback path for the disturbance input and the controller transfer function has a pole there. The zero at $s = -20$ is again due to the pole of the sensor which is in the feedback path for both configurations.

This is what we desire because the system's output $Y(s)$ should follow changes in its reference input $R(s)$ but be unaffected by changes in its disturbance input $D(s)$. By inspection, we can see that the steady-state gain for a reference input is $T_{\text{ref}}(0) = 960/1005 = 0.9552$, which is close to the desired value of 1.0. For the disturbance input the steady-state gain is $T_{\text{dist}}(0) = 180/1005 = 0.1791$, which is close to the desired value of zero. These gains are the steady-state values of the unit step responses.

Figure 3.7 shows the responses of the system to unit-step functions on the two inputs. The upper curve (solid) is the response to the reference input and exhibits a small amount of overshoot before the output settles to its steady-state value of 0.9552, which is slightly less than the desired value of unity. The lower curve (dashed) is the response to the disturbance input, which we would ideally like to be zero for all time. However, it does remain below 0.20

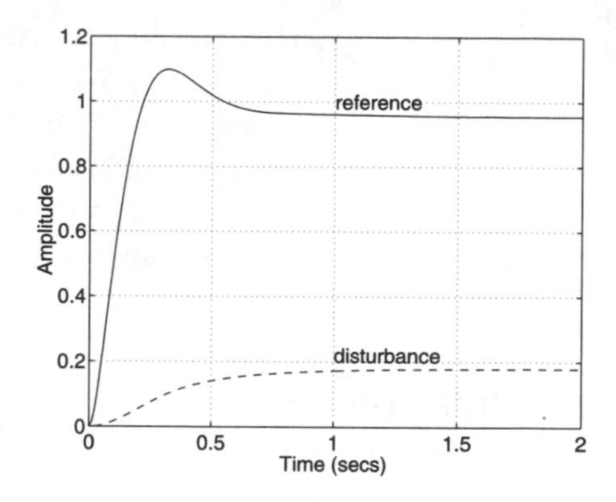

FIGURE 3.7 *Responses of the system in Example 3.5 to steps in the reference and disturbance inputs*

which may or may not be satisfactory, depending on the circumstances of the physical system.

MATLAB Script

```
% Script 3.5:  Feedback system with reference and disturbance inputs
Gc = tf(80*[1 3],[1 15*3])                   % lead controller in TF form
Gp = tf(4,conv([0.5 1],[2 1]))               % plant in TF form
H = tf(1,[0.05 1])                           % sensor in TF form
%------ CL transfer function for reference input
T_ref = feedback(Gp*Gc,H)
%-----CL zeros, poles, & DC gain for reference input
[zT_ref,pT_ref,kT_ref] = zpkdata(T_ref,'v')
T_ref_gainDC = dcgain(T_ref)
t = [0:0.01:2];                          % time vector for 2 seconds
y_ref = step(T_ref,t)                    % response to step in reference input
%
%=========== repeat for disturbance input ===========
T_dist = feedback(Gp,Gc*H)
%-------CL zeros, poles, & DC gain for disturbance input
[zT_dist,pT_dist,kT_dist] = zpkdata(T_dist,'v')
T_dist_gainDC = dcgain(T_dist)
y_dist = step(T_dist,t);                % response to step in disturbance input
plot(t,y_ref,t,y_dist,'--')             % plot step responses on same axes
text(1.0,1.00,'reference input')        % add labels to plots
text(1.0,0.22,'disturbance input')
```

WHAT IF? Replace the gain of 80 in the numerator of $G_c(s)$ in Example 3.5 with the variable gain K whose value you can specify with the `input` command, as in `K = input('enter controller gain....')`.

Then plot the responses to both reference and disturbance step inputs for several values of K. You should find that as the gain is increased: (i) the overshoot increases and the response becomes more oscillatory, (ii) the initial response is faster, and (iii) the steady-state errors are reduced. By trial and error, estimate the maximum value of K for a stable closed-loop system. Also try reducing the gain. How are these features of the response affected? ■

EXPLORATION

E3.1 Interactive analysis with MATLAB. The MATLAB file `one_blk.m` can be used to perform a variety of analytical tasks on any single-block system whose transfer function exists in the workspace as the row vectors `num` and `den`. You can create your own one-block models as described in the statement of Exploratory Problem EP2.1 in Chapter 2.

Alternatively, you can select the entry "Activate keyboard" on the model-building menu. At this point you can (i) enter your own model, (ii) create interconnections of models, or (iii) create a TF model of one of the real-world systems described in Appendix A by running the appropriate M-file and making choices on the menu to select the desired transfer function. At the conclusion of the M-file, the transfer function of the plant or process will be in the workspace. If you rename the vectors that contain the coefficients of the numerator and denominator polynomials of the closed-loop system as `num` and `den`, you can use `one_blk.m` to perform extensive analysis without having to enter more MATLAB commands.

To summarize, you can create models of most any single-input/single-output system you wish and then use `one_blk.m` to do the analysis, provided only that you use the names `num` and `den` for the numerator and denominator coefficient vectors.

COMPREHENSIVE PROBLEMS

CP3.1 Electric power generation system. Consider the voltage control of the electric power system shown in Figure A.2 in Appendix A. Using a lag controller with its transfer function $G_c(s)$ given by (3.3), put the zero of $G_c(s)$ at $s = -1$ and let $\alpha_{lag} = 10$. Vary the controller gain K_{lag} in the range of 10 to 60, and obtain the transfer function of the closed-loop system. Determine the poles, zeros, and DC gain of the closed-loop system, and plot its response to a unit step in the reference voltage input V_{ref}. You will find that in the closed-loop system, the lightly-damped electromechanical mode becomes unstable at a gain of about $K_{\text{lag}} = 50$. The instability can be observed as growing oscillations about the steady-state value of the step response.

CP3.2 Satellite with reference input. Construct a feedback system for controlling the pointing angle of the satellite according to the block diagram in Figure 3.8. In each part of the problem, obtain the transfer function of the closed-loop system with the specified controller and parameter values, determine the poles, zeros, and DC gain of the closed-loop system, and plot its response to a unit step in the desired pointing angle. Because of the choices of the positive senses of the variables that have been used in developing the model, the gain of the controller must be *negative* in order for the closed-loop system to be stable. Experiment with the MATLAB model to convince yourself of this.

a. Use a proportional controller with a negative gain. You should find that the closed-loop system is stable, but very lightly damped. The lower the gain, the lower the frequency of oscillation is, but the rate at which the envelope of the oscillations decays is not affected.
b. Use a lead controller, with $G_c(s)$ given by (3.4) where $K_{\text{lead}} = -0.001$, the zero is at $s = -0.02$, and $\alpha_{\text{lead}} = 20$. With these values you should get a response of the pointing angle to a step change in the desired angle whose maximum value does not exceed its steady-state value by more than 15% and whose steady-state error is zero. Because of the presence of viscous friction between the wheel and the satellite, the steady-state error will be zero with a lead controller only if the steady-state speed of the reaction wheel relative to the satellite is zero. Try some other values for the gain, the zero location, and α_{lead} to see how the response is affected.

CP3.3 Satellite with disturbance input. The block diagram in Figure 3.9 shows how we can create a transfer-function model of the satellite from which we can obtain the response to disturbance inputs with the control system active. Note that there is no input for the disturbance torque in the block diagram of the satellite shown in Figure 3.8, so it cannot be used for this purpose. In each part of the problem, obtain the transfer function of the closed-loop system with the specified controller parameter values, determine the poles, zeros, and DC gain

FIGURE 3.8 *Satellite TF model for pointing angle in response to desired pointing*

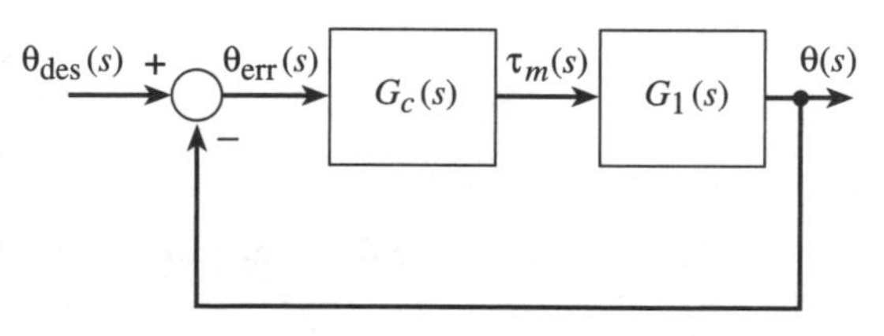

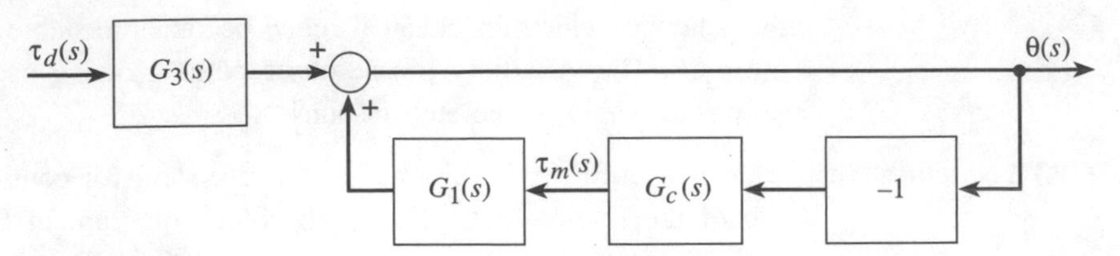

FIGURE 3.9 *Satellite TF model for pointing angle in response to disturbance torque*

of the closed-loop system, and plot its response to a unit impulsive disturbance torque of 0.002 N-m-s.

a. Use a lead controller with the parameters given in part (b) of the previous problem. You should find that following the transients, there will be a steady-state pointing error of approximately 0.3°.
b. Use a lead-lag controller (3.5), with the lag zero being 1/10 of the lead zero. You should find that the steady-state error has been reduced by a factor of $1/\alpha_{\mathrm{lag}}$ compared with the response for the lead controller.

CP3.4 Satellite with wheel speed as output. Construct a transfer-function model of the satellite as shown in Figure 3.10 with a lead controller for $G_c(s)$. Use the transfer function from motor torque to pointing angle for $G_1(s)$ and the transfer function from motor torque to wheel speed as $G_2(s)$. The transfer function for the lead controller is given by (3.4) and the two satellite transfer functions can be obtained by running the file `sat.m` and making the appropriate selections on the menu. Run the model for several choices of the controller parameters, starting with the values given in part (b) of Comprehensive Problem CP3.2. You should find that the wheel speed always starts and finishes at $\Omega = 0$ (this is because the TF model does not allow for nonzero initial conditions) but reaches different maximum or minimum values, depending on the value of the gain K_{lead}.

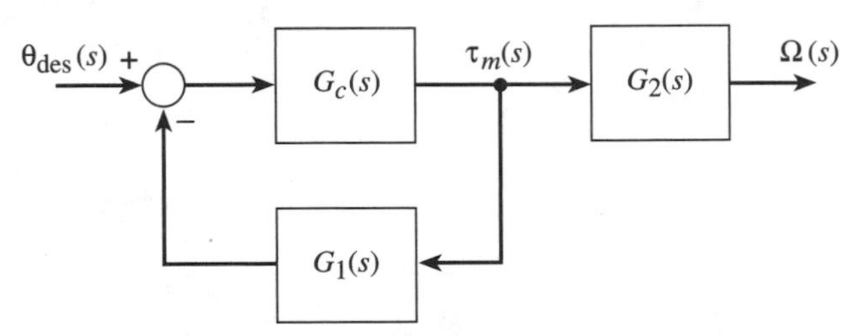

FIGURE 3.10 *Satellite TF model for wheel speed in response to desired pointing angle*

CP3.5 Hydro-turbine system. Construct a feedback system for controlling the mechanical power produced by the turbine according to the block diagram in Figure A.8 in Appendix A. In each case, obtain the transfer function of the closed-loop system with the specified controller and parameter values, determine the poles, zeros, and DC gain of the closed-loop system, and plot its response to a unit step in the desired power.

a. Use a proportional controller with $K_p = 0.5$. You should find that the closed-loop system is stable, but the response reaches a steady-state value of only 0.333, whereas a steady-state response of 1.0 is desired. Also show that the system becomes marginally stable when the gain is increased to 0.75. You should find that the closed-loop poles are $s = \pm j1.871$, which says that the response will contain an undamped oscillation with a period of $2\pi/1.871 = 3.36$ s. Check this on the step response.
b. Replace the proportional controller with a lead controller having its zero at $s = -3$ and $\alpha_{\text{lead}} = 20$. Experiment with the gain to see if you can get a faster response with less overshoot than in part (a). You should find that for a unit-step input the steady-state response is 0.333 when $K_{\text{lead}} = 10$, but the response becomes very oscillatory for larger values of the gain.
c. Try a lead-lag controller with the same values for the lead zero as above and $\alpha_{\text{lead}} = \alpha_{\text{lag}} = 20$, and with the lag zero set to 1/10 of the lead zero ($-3/10 = -0.3$). With $K_{\text{ldlg}} = 10$ the steady-state response will be increased to 0.909 (it was 0.333 without the lag term). Try adjusting the controller parameters to see if you can get less overshoot while retaining a steady-state response of at least 0.90.

SUMMARY

In this chapter we have shown how to combine transfer-function models of two or more subsystems in order to construct more complicated and versatile system models. We have considered series, parallel, and feedback combinations of models that are described by transfer functions in the TF form, namely as a ratio of polynomials. Once the transfer function of the combined system is available, we can perform analytical calculations such as finding the zeros, poles, and gain or establishing the stability of the system. We can also obtain plots of the response to step functions, impulses, or arbitrary inputs. In Chapter 4 we will show how MATLAB can be used to construct models of more complicated systems in state-space form. The Control System Toolbox also has a command named `connect` for building models having a more general structure than the examples we have considered here, but we have not illustrated it.

MATLAB FUNCTIONS USED

Function	*Purpose and Use*	*Toolbox*
*	Given two LTI objects, the * operator forms their series connection.	Control System
+	Given two LTI objects, the + operator forms their parallel connection.	Control System
conv	Given two row vectors containing the coefficients of two polynomials, **conv** returns a row vector containing the coefficients of the product of the two polynomials.	MATLAB
dcgain	Given a model in TF form, **dcgain** returns the steady-state gain of the system.	Control System
feedback	Given the models of two systems, **feedback** returns the model of the closed-loop system, where negative feedback is assumed, unless a third argument is given.	Control System
pole	Computes the poles of an LTI object.	Control System
step	Given a state-space or TF model of a continuous system, **step** returns the response to a unit-step function input.	Control System
text	Adds text to plots at the specified location.	MATLAB
tf	Given numerator and denominator polynomials, **tf** creates the system model as a TF object. The command also converts zero-pole-gain or state-space models to TF form.	Control System
zpk	Given a system's zeros, poles, and gain, **zpk** creates the system model as a ZPK object. The command also converts transfer-function or state-space models to ZPK form.	Control System
zpkdata	Given a ZPK object, **zpkdata** extracts the zeros, poles, and gain and other information about the system.	Control System

ANSWERS

P3.1 Zeros are $s = -0.3333 \pm j0.4714$ and -0.5; poles are $s = -0.6250 \pm j1.2686$ and $-1 \pm j3$; gain is 3. Stable.

P3.2 Zero is $s = -4$; poles are $s = -1$, -2 ± 3.4641, and -8; gain is 10. Stable.

P3.4 Zeros are $s = -0.3333$, -0.5, and -1.5; poles are $s = -0.2 \pm j0.6$, $-1.0 \pm j2.6458$, and -5; gain is 12. Stable.

P3.5 After cancellation, zero is $s = -4$; poles are $s = -2.50 \pm j1.9365$, and -6.0; gain is 8. Stable.

P3.6 After cancellation, zeros are $s = -1.0$ and -1.5; poles are $s = -1.0 \pm j2.0, -2.1$, and -3.0; gain is 1. Unstable.

P3.7 $T(s) = (15s^3 + 93s^2 + 56s + 33)/(30s^3 + 17s^2 + 14s + 2)$

P3.8 $T(s) = (6s^4 + 39s^3 + 128s^2 + 91s + 60)/(10s^4 + 44s^3 + 114s^2 + 156s + 36)$

P3.9 After cancellation, $T(s) = 2(s + 2)/(s + 1)(s + 3)$. Stable.

P3.10 After cancellation, $T(s) = 2(s + 2)/(s + 1)(s + 3)$. Unstable.

P3.11 Zeros are $s = -0.2607 \pm j0.3322, -0.6667, -3.7024$, and $-14.096 \pm j5.789$; poles are $s = -0.3125 \pm j0.3903, -0.3750 \pm j1.1659, -2.0, -10.0$, and -20.0; gain is 0.375. Stable.

P3.12 Zeros are $s = -0.0626, -0.3006$, and $-2.4791 \pm j5.3801$; poles are $s = -0.0625, -0.2087, -0.3000, -0.3820, -2.6180$, and -4.7913; gain is 0.700. Stable.

P3.13 $T(s) = (2s^3 + 19s^2 + 25s + 8)/(4s^4 + 41s^3 + 93s^2 + 130s + 71)$

P3.14 $T(s) = (45s^2 + 273s + 18)/(15s^4 + 61s^3 + 139s^2 + 176s + 23)$

P3.15 $T(s) = (5s^3 + 47s^2 + 99s + 36)/(s^4 + 17.5s^3 + 81.5s^2 + 96.5s + 87)$

P3.16 Zero is $s = -20$; poles are $s = -1.0065$ and $-12.4968 \pm j4.7619$

P3.17 For $K = 1000$, the zero is $s = -8$; poles are $s = +3.7527 \pm j9.0714$ and $-8.7527 \pm j2.5302$. Stable for $K = 1$ and 10; unstable for $K = 100$ and 1000.

P3.18 For $K = 10$, the zero is $s = -5$; poles are $s = -7.0 \pm j2.0$.

P3.19 $T(s) = (4s^3 + 17s^2 + 48s + 11)/(5s^4 + 23s^3 + 65s^2 + 16s + 16)$

CHAPTER 4

State-Space Models

PREVIEW

The use of transfer functions to represent a linear, time-invariant system is suitable when the model is given in the input-output form and when the model order is low. However, models of real systems are usually derived from physical laws in terms of state variables that correspond to identifiable quantities such as stored energies. Such models are put in the state-space form and can be compactly represented in matrix notation. One advantage of the state-space form is that the effects of nonzero initial conditions on the system response can be readily investigated. In addition, for high-order systems, the state-space form is preferable because it is less susceptible to numerical ill-conditioning than the transfer-function form. Control System Toolbox commands are coded to accept the model of a system represented in any of the three forms: transfer-function, zero-pole-gain, or state-space. In this chapter we will apply the MATLAB commands discussed in Chapters 2 and 3 in order to build and analyze state-space models.

MODEL BUILDING, CONVERSIONS, AND INTERCONNECTIONS

For a linear, time-invariant system with n states, m inputs, and p outputs, the state-space model in matrix notation is

$$\dot{\mathbf{x}}(t) = \mathbf{A}\mathbf{x}(t) + \mathbf{B}\mathbf{u}(t), \qquad \mathbf{y}(t) = \mathbf{C}\mathbf{x}(t) + \mathbf{D}\mathbf{u}(t) \tag{4.1}$$

where the state vector $\mathbf{x}(t)$ is of dimension n, the input vector $\mathbf{u}(t)$ of dimension m, and the output vector $\mathbf{y}(t)$ of dimension p. The state matrix $\mathbf{A}$ is of dimension $n \times n$, the input matrix $\mathbf{B}$ is $n \times m$, the output matrix $\mathbf{C}$ is $p \times n$, and the feedforward matrix $\mathbf{D}$ is $p \times m$. For a time-invariant system, $\mathbf{A}$, $\mathbf{B}$, $\mathbf{C}$, and $\mathbf{D}$ are constant matrices. We will frequently refer to (4.1) as the SS form. The development of a state-space model from a physical system can be found in many control systems engineering textbooks. Once the matrices $\mathbf{A}$, $\mathbf{B}$, $\mathbf{C}$, and $\mathbf{D}$ are known, the state-space model can be readily entered in MATLAB by defining the four matrices and using the `ss` command. For example, if we have defined the matrices $\mathbf{A}$, $\mathbf{B}$, $\mathbf{C}$, and $\mathbf{D}$ as the MATLAB variables, `a`, `b`, `c` and `d`, the model can be created as a state-space (SS) object by entering `Gss = ss(a,b,c,d)`.

The state-space model (4.1) has the matrix transfer function $\mathbf{G}(\mathbf{s})$ which obeys the relation

$$\mathbf{G}(s) = \mathbf{C}(s\mathbf{I} - \mathbf{A})^{-1}\mathbf{B} + \mathbf{D}$$

where $\mathbf{I}$ is the identity matrix of the same dimension as $\mathbf{A}$. If a system model `G1` exits in TF form, a state-space equivalent can be obtained by entering `G1ss = ss(G1)`. It should be kept in mind that there is not a unique state-space representation for a given transfer function. In fact, the number of state variables, which is the number of rows or columns of $\mathbf{A}$, may vary from one representation to another. The state-space representation obtained by using the `ss` command is guaranteed to be minimal only if the system is either single-input or single-output.

In similar fashion, if a system model `G2` exists in ZPK form, a state-space equivalent can be obtained by entering `G2ss = ss(G2)`. We will discuss equivalent state-space models in a later section on state-variable transformations.

The Control System Toolbox also provides the function `ssdata` for extracting properties of a SS object. For example, we would enter `[a,b,c,d] = ssdata(S1)` to obtain the four matrices describing the state-space system `S1`. We can also use the `ssdata` command on TF or ZPK objects. In either of these cases, the TF or ZPK object is first converted to SS form.

The commands for interconnecting models that were introduced in Chapter 3 for TF and ZPK systems can be used in exactly the same manner with systems expressed as SS objects. Series connections are made with the * operator, where particular attention must be paid to the ordering of the systems. Parallel connections are made with the + operator, where attention must be paid to the numbers of inputs and outputs when dealing with multi-input/multi-output (MIMO) systems. The `feedback` command allows the interconnection of subsystems in state-space or other forms. Again, care must be taken when working with MIMO systems to insure that the intended input and output pairings are being made.

One additional consideration is that the user must be aware of the precedence rules when there is a mixture of object types. Because the state-space representation is the best of the three in terms of numerical reliability, if at least one of the subsystems being connected is of type SS, the resulting object will be SS. If none of the subsystems is SS, but at least one is of type ZPK, the resulting object will be ZPK. The overall connection will be of type TF only if all of its components are TF objects. Interconnections that differ from these standard forms can be accomplished by using MATLAB to implement the appropriate matrix algebraic equations.

The following two examples illustrate how to enter a SS form of a system, to convert it to the TF form, and to perform interconnections using the SS form.

EXAMPLE 4.1
Building a State-Space Model

Enter in MATLAB the matrices

$$\mathbf{A} = \begin{bmatrix} 0 & 1 & 0 \\ -4 & -1 & 1 \\ 0 & 0 & -20 \end{bmatrix}, \qquad \mathbf{B} = \begin{bmatrix} 0 \\ 0 \\ 20 \end{bmatrix}$$

$$\mathbf{C} = [1 \quad 0 \quad 0], \qquad \mathbf{D} = [0]$$

and create the model as a SS object. Then convert the model to TF form, and then back to SS form, using a different name. Then convert the original SS form to ZPK form, and back to SS form.

Solution

MATLAB allows the entry of a matrix by rows, with the semicolon denoting the end of a row. The entry of the matrices **A**, **B**, **C**, and **D** is shown in Script 4.1.

The SS object named `Gss` is created by the command `Gss = ss(A,B,C,D)` which displays the four matrices, with the columns and rows labeled according to the particular state, input, or output variables involved, using the generic notation `x1`, `x2`, `x3` for the three states, `u1` for the single input, and `y1` for the single output.

The command `Gtf = tf(Gss)` creates the system as a TF object and causes the transfer function to be displayed as the ratio of its numerator and denominator polynomials, namely

$$G(s) = \frac{20}{s^3 + 21s^2 + 24s + 80}$$

The command `Gss1 = ss(Gtf)` converts the TF object back to SS form, with the name `Gss1`, to make it distinguishable from the original SS object `Gss`.

The resulting matrices are

$$\mathbf{A1} = \begin{bmatrix} -21 & -3 & -2.5 \\ 8 & 0 & 0 \\ 0 & 4 & 0 \end{bmatrix}, \qquad \mathbf{B1} = \begin{bmatrix} 1 \\ 0 \\ 0 \end{bmatrix}$$
$$\mathbf{C1} = [0 \quad 0 \quad 0.625], \qquad \mathbf{D1} = [0]$$

The form of **A1** and **B1** is a variant of the more familiar *controller form*. In **A1**, the lower diagonal entries are not ones, as is the case for the controller form. If you compare the state-space matrices of the two representations, you will see that they are not the same. However, by using commands discussed in Example 4.3, the representations can be shown to have the same zeros, poles, and DC gain, which means that the two systems are equivalent from an input-output point of view.

Next, we convert the original SS object `Gss` to ZPK form with the command `Gzpk = zpk(Gss)`, getting the system's transfer function displayed in zero-pole-gain form. Finally, we convert this ZPK form back to SS form by entering `Gss2 = ss(Gzpk)`. The resulting state-space matrices are

$$\mathbf{A2} = \begin{bmatrix} -20 & 0 & 2 \\ 0 & -1 & -2 \\ 0 & 2 & 0 \end{bmatrix}, \qquad \mathbf{B2} = \begin{bmatrix} 0 \\ 2.2361 \\ 0 \end{bmatrix}$$
$$\mathbf{C2} = [2.2361 \quad 0 \quad 0], \qquad \mathbf{D2} = [0]$$

Note that **A2** is in the upper block-triangular form, in which its diagonal blocks are of dimension either 1×1 or 2×2. Clearly, these matrices are quite different from the two other sets. However, they do represent the same transfer function from input to output. As such, they provide a valid state-space representation of the original system.

MATLAB Script

```
% Script 4.1: Entering a state-space model
A = [0 1 0; -4 -1 1; 0 0 -20]          % enter model data
B = [0; 0; 20], C = [1 0 0], D = 0
Gss = ss(A,B,C,D)                      % Set up model in SS form
Gtf = tf(Gss)                          % convert to TF form
Gss1 = ss(Gtf)                         % convert back to SS form
Gzpk = zpk(Gss)                        % convert to ZPK form
Gss2 = ss(Gzpk)                        % convert back to SS form
```

REINFORCEMENT PROBLEMS

For each of the following problems, enter the state matrices of the system into MATLAB and create the model as a SS object. Then convert the model to TF form and then back to SS form, and compare the **A**, **B**, **C**, and **D** matrices of the two forms. Repeat the process using the ZPK form as the intermediate one.

P4.1 Third-order system.

$$\mathbf{A} = \begin{bmatrix} -4 & 1 & 2 \\ 1 & -5 & 3 \\ 2 & 0 & -6 \end{bmatrix} \qquad \mathbf{B} = \begin{bmatrix} 1 \\ 0.5 \\ 2 \end{bmatrix}$$

$$\mathbf{C} = [2 \quad 1 \quad 2] \qquad \mathbf{D} = [0]$$

P4.2 Fourth-order system with $\mathbf{D} \neq 0$.

$$\mathbf{A} = \begin{bmatrix} -2 & 2 & 0 & 0 \\ 0 & -4 & 4 & 0 \\ 0 & 0 & -5 & 5 \\ 0 & 0 & 0 & -10 \end{bmatrix} \qquad \mathbf{B} = \begin{bmatrix} 0 \\ 0 \\ 0 \\ 10 \end{bmatrix}$$

$$\mathbf{C} = [1 \quad 0 \quad 1 \quad 0] \qquad \mathbf{D} = [0.5]$$

EXAMPLE 4.2 ***Series, Parallel, and Feedback Connections***

For the systems with transfer functions $G_1(s)$ and $G_2(s)$ given by the state-space models

$$\mathbf{A}_1 = \begin{bmatrix} 0 & 1 \\ -3 & -5 \end{bmatrix}, \qquad \mathbf{B}_1 = \begin{bmatrix} 0 \\ 1 \end{bmatrix}, \qquad \mathbf{C}_1 = [1 \quad 2], \qquad \mathbf{D}_1 = [0]$$

$$\mathbf{A}_2 = \begin{bmatrix} -3 & 1 \\ 0 & -4 \end{bmatrix}, \qquad \mathbf{B}_2 = \begin{bmatrix} 0 \\ 4 \end{bmatrix}, \qquad \mathbf{C}_2 = [3 \quad 0], \qquad \mathbf{D}_2 = [2]$$

find state-space models for the following interconnections:

a. series: $T_s(s) = G_1(s)G_2(s)$
b. parallel: $T_p(s) = G_1(s) + G_2(s)$
c. feedback:

$$T_f(s) = \frac{G_1(s)}{1 + G_1(s)G_2(s)}$$

Solution The MATLAB commands in Script 4.2 will compute the series, parallel, and feedback connections of $G_1(s)$ and $G_2(s)$ expressed in the SS form.

MATLAB Script

```
% Script 4.2: Interconnections in state-space form
A1 = [0 1; -3 -5], B1 = [0; 1]        % enter G1 state—space matrices
C1 = [1 2], D1 = 0
G1 = ss(A1,B1,C1,D1)                  % build G1(s) as SS object
A2 = [-3 1; 0 -4], B2 = [0; 4]        % enter G2 state—space matrices
C2 = [3 0], D2 = 2
G2 = ss(A2,B2,C2,D2)                  % build G1(s) as SS object
Ts = G1*G2                            % Series connection Ts(s)
Tp = G1+G2                            % Parallel connection Tp(s)
Tf = feedback(G1,G2)                  % Feedback connection Tf(s)
```

The series connection results in the matrices

$$\mathbf{A}_s = \begin{bmatrix} 0 & 1 & 0 & 0 \\ -3 & -5 & 3 & 0 \\ 0 & 0 & -3 & 1 \\ 0 & 0 & 0 & -4 \end{bmatrix} \qquad \mathbf{B}_s = \begin{bmatrix} 0 \\ 2 \\ 0 \\ 4 \end{bmatrix}$$

$$\mathbf{C}_s = [1 \quad 2 \quad 0 \quad 0] \qquad \mathbf{D}_s = [0]$$

The parallel connection results in

$$\mathbf{A}_p = \begin{bmatrix} 0 & 1 & 0 & 0 \\ -3 & -5 & 0 & 0 \\ 0 & 0 & -3 & 1 \\ 0 & 0 & 0 & -4 \end{bmatrix} \qquad \mathbf{B}_p = \begin{bmatrix} 0 \\ 1 \\ 0 \\ 4 \end{bmatrix}$$

$$\mathbf{C}_p = [1 \quad 2 \quad 3 \quad 0] \qquad \mathbf{D}_p = [2]$$

Note the special structures of these state matrices. For example, the main 2×2 diagonal blocks of $\mathbf{A}_s$ and $\mathbf{A}_p$ are $\mathbf{A}_1$ and $\mathbf{A}_2$. Also verify that the upper 2×2 off-diagonal block of $\mathbf{A}_s$ is $\mathbf{B}_1\mathbf{C}_2$, and $\mathbf{D}_p = \mathbf{D}_1 + \mathbf{D}_2$.

The feedback connection results in the closed-loop system matrices

$$\mathbf{A}_f = \begin{bmatrix} 0 & 1 & 0 & 0 \\ -5 & -9 & -3 & 0 \\ 0 & 0 & -3 & 1 \\ 4 & 8 & 0 & -4 \end{bmatrix} \qquad \mathbf{B}_f = \begin{bmatrix} 0 \\ 1 \\ 0 \\ 0 \end{bmatrix}$$

$$\mathbf{C}_f = [1 \quad 2 \quad 0 \quad 0] \qquad \mathbf{D}_f = [0]$$

Because of the presence of feedback, the structure of $\mathbf{A}_f$ is more complicated than that of its counterparts $\mathbf{A}_s$ and $\mathbf{A}_p$.

REINFORCEMENT PROBLEMS

In Problems 4.3 and 4.4, find the SS forms of the series, parallel, and feedback connections of the systems $G_1(s)$ and $G_2(s)$.

P4.3 $G_1(s)$ and $G_2(s)$ given in SS form.

$$\mathbf{A}_1 = \begin{bmatrix} 0 & 1 & 0 \\ 0 & 0 & 1 \\ -16 & -20 & -5 \end{bmatrix}, \qquad \mathbf{B}_1 = \begin{bmatrix} 0 \\ 0 \\ 1 \end{bmatrix}$$

$$\mathbf{C}_1 = [8 \quad 2 \quad 0], \qquad \mathbf{D}_1 = [0]$$

$$\mathbf{A}_2 = [-8], \qquad \mathbf{B}_2 = [1], \qquad \mathbf{C}_2 = [5], \qquad \mathbf{D}_2 = [0]$$

P4.4 $G_1(s)$ and $G_2(s)$ given in TF form.

$$G_1(s) = \frac{3s^2 + 2s + 1}{4s^2 + 5s + 8} \quad \text{and} \quad G_2(s) = \frac{4s + 2}{s^2 + 2s + 10}$$

Do this problem by first converting $G_1(s)$ and $G_2(s)$ to SS form and then performing the interconnections in SS form.

P4.5 Feedback connection of $G_1(s)$ and $G_2(s)$. Redo the feedback connection of the systems in Problem 4.4 by first performing the feedback connection in TF form and then converting the resulting system to SS form. You will obtain a state-space model with matrices **A**, **B**, and **C** that are different from those of the corresponding model obtained in Problem 4.4. In Problem 4.8, you will be asked to show that these two models are equivalent in the sense that they both have the same transfer function.

POLES, ZEROS, EIGENVALUES, AND STABILITY

The poles and zeros of a single-input/single-output (SISO) linear, time-invariant system in state-space form (4.1) can be computed directly from the SS form. This is done by using the Control System Toolbox command `[z,p,k] = zpkdata(G,'v')`, where the second input argument `'v'` forces the outputs `z` and `p` to appear as column vectors. The column vectors `z` and `p` contain the zeros and poles of the transfer function, respectively, and the scalar `k` contains the gain. For a multi-input system, the form of the command is `[zz,pp,kk] = zpkdata(G)` and the outputs `zz` and `pp` are *cell arrays*, and `kk` is a matrix.

Alternatively, the poles can be computed by using the command `pole(G)`. If all the eigenvalues of **A** are in the open left half-plane, the system is asymptotically stable. For a SISO system, the zeros obtained from the ZPK form can also be computed by the transmission-zero command `tzero(G)` from the Control System Toolbox. A plot of the poles and zeros can be obtained directly from the command `pzmap(G)`. For high-order systems, the computation of the poles and zeros from the state-space matrices is numerically more reliable than solving for the roots of the numerator and denominator polynomials of a transfer function. The following example illustrates the use of the functions `ss`, `zpkdata`, `pole`, `tzero`, and `pzmap`.

EXAMPLE 4.3
Eigenvalues and Transmission Zeros

For the system whose state-space representation is

$$\mathbf{A} = \begin{bmatrix} 0 & 1 & 0 \\ -5 & -1 & 1 \\ 0 & 0 & -10 \end{bmatrix}, \qquad \mathbf{B} = \begin{bmatrix} 0 \\ 0 \\ 10 \end{bmatrix}$$

$$\mathbf{C} = [1 \quad 0 \quad 1], \qquad \mathbf{D} = [0]$$

obtain the zeros, poles, and gain of its transfer function. Verify the poles and zeros of the system using the functions `pole` and `tzero`. Plot the zeros and poles on the s-plane. Comment on the stability of the system.

Solution

Following the MATLAB commands in Script 4.3, the conversion of the state-space model to ZPK form yields the zeros at $s = -0.5 \pm j2.398$, the poles at $s = -0.5 \pm j2.179$ and -10, and a gain of 10. The pole and the transmission-zero computations yield the same results. Since all the poles of the system are in the left half-plane, the system is asymptotically stable. To plot the poles and zeros, we use the function `pzmap` with the SS object `G` as the single argument.

MATLAB Script

```
% Script 4.3: Eigenvalues and transmission zeros
A = [0 1 0; -5 -1 1; 0 0 -10]          % enter state-space matrices
B = [0; 0; 10], C = [1 0 1], D = 0
G = ss(A,B,C,D)                         % build system as SS object
[zG,pG,kG] = zpkdata(G,'v')             % Zeros, poles and gain
poleG = pole(G)                         % System poles
zeroG = tzero(G)                        % System transmission zeros
pzmap(G)                                % draw pole-zero plot
axis([-12 0 -3 3])                      % expand plotted area
```

REINFORCEMENT PROBLEMS

In Problems 4.6 and 4.7, convert the state-space model to its transfer function in ZPK form; that is, find its zeros, poles, and gain. Then use the functions `pole` and `tzero` to verify the poles and zeros obtained by using the `zpkdata` command. Plot the poles and zeros on the s-plane.

P4.6 Third-order system. The state-space model of Problem 4.1.

P4.7 Fourth-order system with D≠ 0. The state-space model of Problem 4.2. Since **A** is upper-triangular, the eigenvalues of the **A** matrix are the entries on its main diagonal.

P4.8 Equivalence of state-space models. Use `zpkdata` to find the zeros, poles, and gain of the state-space models obtained from the feedback connections in Problems 4.4 and 4.5. You should find that they are equal, showing that the systems are equivalent even though their state matrices differ.

TIME RESPONSE

One advantage of a state-space model over a transfer-function model is that the initial values of the state variables can be directly included in the model as the nonzero initial condition vector $\mathbf{x}(t_0)$, where t_0 is the initial time. In contrast, we are implicitly assuming zero-input conditions whenever we use either the transfer-function (TF) form or the zero-pole-gain (ZPK) form. The complete response of the state vector $\mathbf{x}(t)$ of the system represented by (4.1) with an arbitrary input $\mathbf{u}(t)$ and the initial condition $\mathbf{x}(t_0)$ is

$$\mathbf{x}(t) = \epsilon^{\mathbf{A}(t-t_0)}\mathbf{x}(t_0) + \int_{t_0}^{t} \epsilon^{\mathbf{A}(t-\tau)}\mathbf{B}\mathbf{u}(\tau)\,d\tau$$

where $\epsilon^{\mathbf{A}t}$ is called the *state transition matrix* and τ is a dummy variable of integration.

The response consists of two parts. In the first part the matrix exponential function $\epsilon^{\mathbf{A}(t-t_0)}$ maps the initial state $\mathbf{x}(t_0)$ to the current state $\mathbf{x}(t)$ if the input $\mathbf{u}(t)$ is zero. The second part of the response is the convolution of the state transition matrix $\epsilon^{\mathbf{A}t}$ with the input $\mathbf{u}(t)$, which requires an integration of a matrix expression for its solution.

In Chapter 2 the commands `impulse`, `step`, and `lsim` were used to simulate the time response of TF and ZPK objects. These commands can also be used for the simulation of state-space (SS) objects. In particular, `lsim` allows the inclusion of a nonzero initial condition as an argument and the calculation of the states as an output variable. For example, the command `[y,t,x] = lsim(G,u,t,x0)` requires `t` to be a column vector of uniformly spaced time values, the input `u` to have as many columns as the number of inputs and as many rows as `t`, and `x0` to be a column vector of the initial conditions of the n state variables. The output response is in `y`. The state response is returned in `x`, which has n columns and as many rows as `t`.

In the following example, we illustrate the use of the time-simulation functions `impulse`, `step`, and `lsim` for a state-space object.

EXAMPLE 4.4
Impulse, Step, and General Response

Use MATLAB to compute and plot the impulse and step responses of the state-space model

$$\mathbf{A} = \begin{bmatrix} -2 & -2.5 & -0.5 \\ 1 & 0 & 0 \\ 0 & 1 & 0 \end{bmatrix}, \qquad \mathbf{B} = \begin{bmatrix} 1 \\ 0 \\ 0 \end{bmatrix}$$

$$\mathbf{C} = [0 \quad 1.5 \quad 1], \qquad \mathbf{D} = [0]$$

Then simulate and plot the state response $\mathbf{x}(t) = [x_1(t) \quad x_2(t) \quad x_3(t)]^T$ when the input is

$$u(t) = \begin{cases} 2 & 0 \le t < 2 \\ 0.5 & t \ge 2 \end{cases}$$

and the initial condition is $\mathbf{x}(0) = [1 \quad 0 \quad 2]^T$.

Solution

The MATLAB commands in Script 4.4 are used to simulate the desired responses. The impulse and step responses, which assume zero initial conditions, are identical to those in Figures 2.1 and 2.2. The state response $\mathbf{x}(t)$ due to the input $\mathbf{u}(t)$ and the initial condition $\mathbf{x}(0)$ is shown in Figure 4.1.

MATLAB Script

```
% Script 4.4: Impulse, step, and general response
A = [-2 -2.5 -0.5; 1 0 0; 0 1 0]          % state-space matrices
B = [1; 0; 0], C = [0 1.5 1], D = 0
G = ss(A,B,C,D)                           % build system as SS object
t = [0:0.1:20]';                          % time as column vector
impulse(G,t)                              % generate impulse response
step(G,t)                                 % generate step response
x0 = [1; 0; 2];                           % nonzero initial condition (IC)
u(1:20,1) = 2*ones(20,1);                 % input for 0<t<2
u(21:201) = 0.5*ones(181,1);              % input for t>2
[y,t,x] = lsim(G,u,t,x0);                 % response due to input IC
plot(t,x(:,1),'-',t,x(:,2),'--',t,x(:,3),'-.')  % plot with different lines
text(6,0.3,'x_1(t)')                      % label curves, using problem
text(6,-0.5,'x_2(t)')                     % .....coordinates, and
text(8,1.8,'x_3(t)')                      % .....use '_' for subscripts
```

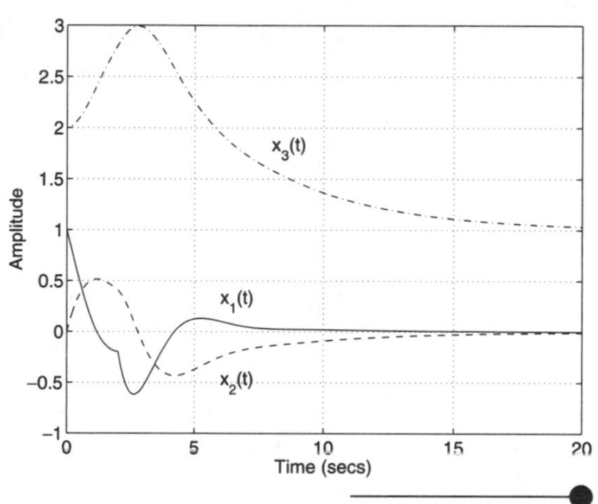

FIGURE 4.1 *Pulse response with nonzero initial condition for Example 4.4: $x_1(t)$, $x_2(t)$, and $x_3(t)$*

REINFORCEMENT PROBLEMS

P4.9 Third-order system. Simulate and plot the impulse and step responses of the state-space model

$$\mathbf{A} = \begin{bmatrix} -7 & -32 & -60 \\ 1 & 0 & 0 \\ 0 & 1 & 0 \end{bmatrix}, \qquad \mathbf{B} = \begin{bmatrix} 1 \\ 0 \\ 0 \end{bmatrix}$$

$$\mathbf{C} = [2 \quad 11 \quad 5], \qquad \mathbf{D} = [0]$$

This is the *controller form* of the system in Problem 2.5.

P4.10 General response. Simulate and plot the response $\mathbf{x}(t)$ and $y(t)$ of the system in Problem 4.9 with a unit-step input $u(t)$ and the initial condition $\mathbf{x}(0) = [2 \quad 0 \quad 1]^T$.

P4.11 A special initial-condition response. Simulate and plot the response $y(t)$ of the system in Problem 4.9 with $u(t) = 0$ and the initial condition $\mathbf{x}(0) = [1 \quad 0 \quad 0]^T$. Note that this response is identical to the impulse response found in Problem 4.9. This problem illustrates that the impulse instantaneously forces the initial condition $\mathbf{x}(0)$ from zero to the nonzero value used here.

STATE TRANSFORMATION

To reveal additional system properties, the state-space model (4.1) can be transformed to a new set of state variables represented by the vector $\mathbf{z}$. Let $\mathbf{T}$ be the nonsingular transformation matrix relating $\mathbf{x}(t) = \mathbf{T}\mathbf{z}(t)$. In the new state vector $\mathbf{z}(t)$, the model (4.1) becomes

$$\dot{\mathbf{z}}(t) = \mathbf{T}^{-1}\mathbf{A}\mathbf{T}\mathbf{z}(t) + \mathbf{T}^{-1}\mathbf{B}\mathbf{u}(t), \qquad \mathbf{y}(t) = \mathbf{C}\mathbf{T}\mathbf{z}(t) + \mathbf{D}\mathbf{u}(t) \tag{4.2}$$

The models (4.1) and (4.2) are equivalent in the sense that they have the same input-output transfer function $\mathbf{G}(s)$ from $\mathbf{U}(s)$ to $\mathbf{Y}(s)$.

A particularly useful transformation is when $\mathbf{T} = \mathbf{M}$, a matrix whose columns are the eigenvectors of $\mathbf{A}$. When $\mathbf{A}$ has distinct eigenvalues, $\mathbf{A}_d = \mathbf{M}^{-1}\mathbf{A}\mathbf{M}$ is a diagonal matrix, and the resulting model (4.2) is said to be in the *modal form*. The diagonal matrix $\mathbf{A}_d$ and the $\mathbf{M}$ matrix can be obtained using the MATLAB eigenvalue command `[M,Ad] = eig(A)` with two output variables. The ith column of $\mathbf{M}$ is the eigenvector corresponding to the ith eigenvalue of the diagonal matrix $\mathbf{A}_d$. The modal form is of interest because it has a parallel structure of first-order subsystems, and its response is in terms of the mode functions of the system. The Control System Toolbox function `canon` also provides an option to find the modal form directly, as shown in the following example.

EXAMPLE 4.5 *Modal Transformation*

Find the eigenvalues and eigenvectors of $\mathbf{A}$ and the modal form of the state-space model of Example 4.4.

Solution The MATLAB commands that will perform the computations required in this example are given in Script 4.5. After entering the state matrices, the command `[M,Ad] = eig(A)` yields the eigenvector matrix

$$\mathbf{M} = \begin{bmatrix} -0.9707 & 0.2624 + j0.2579 & 0.2624 - j0.2579 \\ 0.2337 & -0.5252 + j0.0726 & -0.5252 - j0.0726 \\ -0.0563 & 0.3791 - j0.6632 & 0.3791 + j0.6632 \end{bmatrix}$$

and the diagonal matrix

$$\mathbf{A}_d = \begin{bmatrix} -0.2408 & 0 & 0 \\ 0 & -0.8796 + j1.1414 & 0 \\ 0 & 0 & -0.8796 - j1.1414 \end{bmatrix}$$

Note that $\mathbf{A}_d$ has a conjugate pair of complex eigenvalues, and as a result the corresponding eigenvectors in $\mathbf{M}$ are also complex. The transformed input matrix $\mathbf{B}_d = \mathbf{M}^{-1}\mathbf{B}$ and output matrix $\mathbf{C}_d = \mathbf{CM}$

$$\mathbf{B}_d = [-0.6022 \quad -0.2548 + j0.8740 \quad -0.2548 - j0.8740]^T$$

$$\mathbf{C}_d = [-1.3213 \quad 1.0501 + j0.1491 \quad 1.0501 - j0.1491]$$

are also complex. When we apply the function `canon`, in which the second argument specifies the modal transformation, we obtain the *real* modal form (no complex numbers) of the system as

$$\mathbf{A}_m = \begin{bmatrix} -0.2408 & 0 & 0 \\ 0 & -0.8796 & 1.1414 \\ 0 & -1.1414 & -0.8796 \end{bmatrix}, \qquad \mathbf{B}_m = \begin{bmatrix} -0.6022 \\ -0.5095 \\ -1.7480 \end{bmatrix}$$

$$\mathbf{C}_m = [-1.3213 \quad 1.0501 \quad 0.1491], \qquad \mathbf{D}_m = [0]$$

When the matrix $\mathbf{A}$ has complex eigenvalues, the `canon` function will implement the modal form with real numbers only. The resulting $\mathbf{A}_m$ will have 2×2 diagonal blocks, instead of complex eigenvalues on the diagonals. The function `eig` can be applied to the lower 2×2 diagonal block of $\mathbf{A}_m$ to show that its eigenvalues are $-0.8796 \pm j1.1414$.

MATLAB Script

```
% Script 4.5: Modal transformation
A = [0 1 0; 0 0 1; -0.5 -2.5 -2]   % state-space matrices
B = [0; 0; 1],
C = [1 -1.5 0], D = 0
[M,Ad] = eig(A)                    % eigenvalues and eigenvectors
Bd = inv(M)*B                      % transformed input matrix
Cd = C*M                           % transformed output matrix
G = ss(A,B,C,D)                    % build system as SS object
Gm = canon(G,'modal')              % create real modal form as SS object
```

REINFORCEMENT PROBLEMS

For the state-space models given in the following problems, find the eigenvalues and eigenvectors of **A** and the modal form of the system. If the eigenvalues are complex, use the command `canon` to express the system in its real modal form.

P4.12 Third-order system. The state-space model of Problem 4.1.

P4.13 Fourth-order system. The state-space model of Problem 4.2.

EXPLORATION

EP4.1 Interactive analysis with MATLAB. The MATLAB file `one_blk.m` can be used to perform a variety of analytical tasks on a single-block system implemented as an SS object. You can create MATLAB models of the real-world systems described in Appendix A by running the appropriate M-files and making choices on the menus to select the state-space model. At the conclusion of the M-file, the state-space matrices will be in the workspace, with names like `A`, `B`, `C`, and `D`. You can find the eigenvalues and zeros of the model and examine time responses due to an impulse, a step, and a general input. You can create the transfer function of a proportional, lead, lag, or lead-lag controller and convert it to a state-space model. Then use the $*$ operator and `feedback` command to build a closed-loop state-space model in the form of Figure 3.5. The `one_blk` program can then be used to analyze the resulting closed-loop system.

CHAPTER 4

COMPREHENSIVE PROBLEMS

CP4.1 Electric power generation system. Run the file `epow.m` to obtain the state-space model from the input (u) to the three outputs (V_{term}, ω, and P). Use the function `pole` to find the poles of the system and the function `tzero` to find the zeros for each output.

Consider the voltage control of the electric power system shown in Figure A.2 in Appendix A using a lag controller $G_c(s)$ (3.2). Put the zero of the lag at $s = -1$ and use $\alpha_{lag} = 10$. Vary the gain K_{lag} in the range of 10 to 60 and obtain $G_c(s)$ in SS form. For each value of K_{lag}, determine the poles, zeros, and DC gain of the closed-loop system, and plot its response to a unit step in the reference voltage input V_{ref}.

CP4.2 Satellite with reference input. Run the file sat.m to obtain the state-space model of the satellite discussed in Appendix A. The model described by the matrices A, B, C, and D has one input (motor torque), two outputs [pointing angle in degrees and wheel speed in revolutions per minute (rpm)], and three state variables. Find the poles and the zeros from the motor torque to the pointing angle.

Then use the $*$ operator to put a lead controller $G_c(s)$ (3.3) in series with the satellite, so the output of the lead is connected to the motor-torque input of the satellite, as shown in Figure A.4 in Appendix A. A set of controller parameters that will yield a good response when the loop is closed is $K_{\text{lead}} = -0.001$, $z_{\text{lead}} = -0.02$, and $\alpha_{\text{lead}} = 10$. Determine the poles and zeros of the open-loop series combination.

Using the function feedback with the arguments defined so the first output (pointing angle) is connected to the input of the controller with a negative sign at the summing junction, obtain the state-space model of the closed-loop system diagrammed in Figure A.4. See the file cp4_2.m for the details of how to do this. The input is the reference pointing angle in degrees and the outputs are the pointing angle in degrees and the speed of the reaction wheel in rpm. Then use the model to determine the eigenvalues and zeros of the closed-loop system and to plot both outputs following a unit step change in the reference pointing angle.

Experiment with different values of the controller gain and the lead zero to get a feeling for their effects on the stability, response time, maximum and minimum response values, and steady-state behavior. For example, you should find that increasing the magnitude of the gain K_{lead} results in faster response but requires larger excursions in the wheel speed. Show that if $K_{\text{lead}} > 0$ the closed-loop system will be unstable. Can you give a heuristic argument as to why this should be the case?

Finally, use the functions lsim and subplot to compute and plot the responses of both the pointing angle and the wheel speed when the initial wheel speed is nonzero. You should find that this condition will introduce a steady-state error in the pointing angle that is proportional to the initial wheel speed. Can you explain why this happens based on the modeling equations given in Appendix A?

CP4.3 Stick balancer with rigid stick. Run the file rigid.m to generate the four matrices of the state-space model of the rigid stick balancer in the form of (4.1) and verify that the dimensions of each of the matrices are consistent with the number of state variables (4), the number of outputs (2), and the fact that the model has one input (the voltage applied to the motor that drives the cart). Then find the eigenvalues of the open-loop system (you should find one in the right half-plane, one at the origin of the s-plane, and the other two in the left half-plane).

Use the function `lsim` with nonzero initial conditions to simulate the zero-input response to a *positive* initial stick angle of $\theta(0) = 0.1$ rad and a *negative* initial angular velocity $\omega(0) = \omega_0$. Run the simulation a number of times with different values of ω_0 in order to find the initial angular velocity ω_0^* that will cause the stick to approach the vertical position without going over the top or falling back. Because this is an unstable equilibrium condition you will not be able to balance the stick for an indefinite period of time, but you should be able to satisfy the condition $|\theta(4)| < 0.005$ rad after a few attempts.

Use the function `canon` to transform the state-space model to modal form. Identify the eigenvalues in the modal **A** matrix and compare them with those found from the original **A** matrix with the function `eig`.

CP4.4 Stick balancer with flexible stick. Using the file `flex.m`, repeat each of the steps in Problem CP4.3 for the stick balancer with a flexible stick. Note that this model has two additional state variables for the bending mode and an additional output that is the normalized displacement of the bending mode.

SUMMARY

In this chapter we have shown how to create and work with state-space models in MATLAB. The computation of the poles and zeros of state-space models was illustrated. The simulation and interconnection functions introduced in Chapters 2 and 3 for transfer-function models were extended to state-space models. We also examined the transformation of a state-space model to its modal form. In Chapter 10 we will show how MATLAB can be used to design controllers for state-space models.

MATLAB FUNCTIONS USED

Function	*Purpose and Use*	*Toolbox*
*	Given two LTI objects, the * operator forms their series connection.	Control System
+	Given two LTI objects, the + operator forms their parallel connection.	Control System
canon	Given a system in state-space form, **canon** returns its modal form.	Control System
eig	Given a square matrix, **eig** computes its eigenvalues and eigenvectors.	MATLAB

feedback	Given the models of two systems, **feedback** returns the model of the closed-loop system, where negative feedback is assumed unless a third argument is given.	Control System
impulse	Given a continuous system, **impulse** returns the response to a unit-impulse input.	Control System
inv	Given a square matrix, **inv** returns its inverse.	MATLAB
lsim	Given a state-space model of a continuous system, an array of input values, a vector of time points, and a vector of initial conditions, **lsim** returns the time response of the output and the state.	Control System
ones	Creates an array of specified size, all of whose elements are unity.	MATLAB
pole	Computes the poles of an LTI object.	Control System
pzmap	Given the TF or SS form of a system, **pzmap** produces a plot of the system's poles and zeros in the s-plane.	Control System
ss	Given a set of state-space matrices, **ss** creates the model as an SS object.	Control System
step	Given a continuous system, **step** returns the response to a unit-step input.	Control System
text	Adds text to plots at the specified location.	MATLAB
tf	Given numerator and denominator polynomials, **tf** creates the system model as a TF object. The command also converts zero-pole-gain or state-space models to TF form.	Control System
tzero	Given a state-space model, **tzero** returns the zeros of its transfer function.	Control System
zpk	Given a system's zeros, poles, and gain, **zpk** creates the system model as a ZPK object. The command also converts transfer-function or state-space models to ZPK form.	Control System
zpkdata	Given a ZPK object, **zpkdata** extracts the zeros, poles, and gain and other information about the system.	Control System

ANSWERS

P4.1 Transfer function is $G(s) = (6.5s^2 + 83s + 266)/(s^3 + 15s^2 + 69s + 88)$. State-space form is

$$\mathbf{A}_c = \begin{bmatrix} -15 & -4.313 & -1.375 \\ 16 & 0 & 0 \\ 0 & 4 & 0 \end{bmatrix}, \qquad \mathbf{B}_c = \begin{bmatrix} 4 \\ 0 \\ 0 \end{bmatrix}$$

$$\mathbf{C}_c = [1.625 \quad 1.297 \quad 1.039], \qquad \mathbf{D}_c = [0]$$

P4.2 Transfer function is $G(s) = (0.5s^4 + 10.5s^3 + 124s^2 + 510s + 1000)/(s^4 + 21s^3 + 148s^2 + 420s + 400)$. State-space form is

$$\mathbf{A}_c = \begin{bmatrix} -21 & -9.25 & -3.28 & -1.56 \\ 16 & 0 & 0 & 0 \\ 0 & 8 & 0 & 0 \\ 0 & 0 & 2 & 0 \end{bmatrix}, \qquad \mathbf{B}_c = \begin{bmatrix} 2 \\ 0 \\ 0 \\ 0 \end{bmatrix}$$

$$\mathbf{C}_c = [0 \quad 1.56 \quad 1.17 \quad 1.56], \qquad \mathbf{D}_c = [0.5]$$

P4.3 Feedback connection:

$$\mathbf{A}_f = \begin{bmatrix} 0 & 1 & 0 & 0 \\ 0 & 0 & 1 & 0 \\ -16 & -20 & -5 & -5 \\ 8 & 2 & 0 & -8 \end{bmatrix}, \qquad \mathbf{B}_f = \begin{bmatrix} 0 \\ 0 \\ 1 \\ 0 \end{bmatrix}$$

$$\mathbf{C}_f = [8 \quad 2 \quad 0 \quad 0], \qquad \mathbf{D}_f = [0]$$

P4.4 Feedback connection:

$$\mathbf{A}_f = \begin{bmatrix} -1.25 & -1 & -2 & -0.25 \\ 2 & 0 & 0 & 0 \\ -0.875 & -1.25 & -5 & -2.875 \\ 0 & 0 & 1 & 0 \end{bmatrix}, \qquad \mathbf{B}_f = \begin{bmatrix} 1 \\ 0 \\ 1.5 \\ 0 \end{bmatrix}$$

$$\mathbf{C}_f = [-0.4375 \quad -0.625 \quad -1.5 \quad -0.1875], \qquad \mathbf{D}_f = [0.75]$$

P4.5 Feedback connection:

$$\mathbf{A}_f = \begin{bmatrix} -6.25 & -2.25 & -1.156 & -1.281 \\ 8 & 0 & 0 & 0 \\ 0 & 2 & 0 & 0 \\ 0 & 0 & 1 & 0 \end{bmatrix}, \qquad \mathbf{B}_f = \begin{bmatrix} 2 \\ 0 \\ 0 \\ 0 \end{bmatrix}$$

$$\mathbf{C}_f = [-1.344 \quad -0.2969 \quad -0.2617 \quad -0.4023], \qquad \mathbf{D}_f = [0.75]$$

P4.6 Zeros are at $s = -6.385 \pm j0.3997$; poles are at $s = -2.099$ and $-6.450 \pm j0.5568$; and gain is 6.5.

P4.7 Zeros are at $s = -7.799 \pm j9.467$ and $-2.701 \pm j2.449$; poles are at $s = -2, -4, -5,$ and -10; and gain is 0.5.

P4.8 Zeros are at $s = -0.333 \pm j0.471$ and $-1 \pm j3$; poles are at $s = -0.380 \pm j1.243$ and $-2.745 \pm j2.146$; and gain is 0.75.

P4.12 Diagonal form of the system matrix is

$$\mathbf{A}_d = \text{diag}(-2.0994, -6.4503 + j0.5568, -6.4503 - j0.5568)$$

Real modal form is

$$\mathbf{A}_m = \begin{bmatrix} -2.0994 & 0 & 0 \\ 0 & -6.4503 & 0.5568 \\ 0 & -0.5568 & -6.4503 \end{bmatrix}, \qquad \mathbf{B}_m = \begin{bmatrix} 2.2850 \\ 3.5325 \\ -0.1141 \end{bmatrix}$$

$$\mathbf{C}_m = [2.7386 \quad 0.0652 \quad -0.1042], \qquad \mathbf{D}_m = [0]$$

P4.13 Diagonal form of the system matrix is

$$\mathbf{A}_d = \text{diag}(-2, -4, -5, -10)$$

Modal form is

$$\mathbf{A}_m = \mathbf{A}_d, \qquad \mathbf{B}_m = \begin{bmatrix} 8.333 \\ 47.141 \\ 49.102 \\ 15.723 \end{bmatrix}$$

$$\mathbf{C}_m = [1 \quad -0.7071 \quad 0.7467 \quad -0.7420], \qquad \mathbf{D}_m = [0.5]$$

STATE-SPACE MODELS

CHAPTER 5

Root-Locus Plots

PREVIEW

To design a control system involving the feedback of measured variables to the controller, it is essential that we be able to analyze the *open-loop* system to determine how the *closed-loop* system will behave. One of the most useful techniques for doing this is to construct and interpret the root locus for the system. The root-locus is a plot in the *s*-plane of all possible locations that the roots of a closed-loop system's characteristic equation can have as a specific parameter, usually a gain, is varied from zero to infinity. By examining such a plot, the designer can make intelligent choices for the structure and parameter values of the controller and can infer a good deal of useful information about the performance of the controlled closed-loop system.

DRAWING ROOT-LOCUS PLOTS

We will assume that the feedback system under discussion can be represented as shown in Figure 5.1, where $G(s)$ is the transfer function of the forward path and $H(s)$ is the transfer function of the feedback path. Usually the feedback signal will enter the summing junction with a negative sign, so we will assume this to be the case unless stated to the contrary.

FIGURE 5.1 *Block diagram of a feedback system*

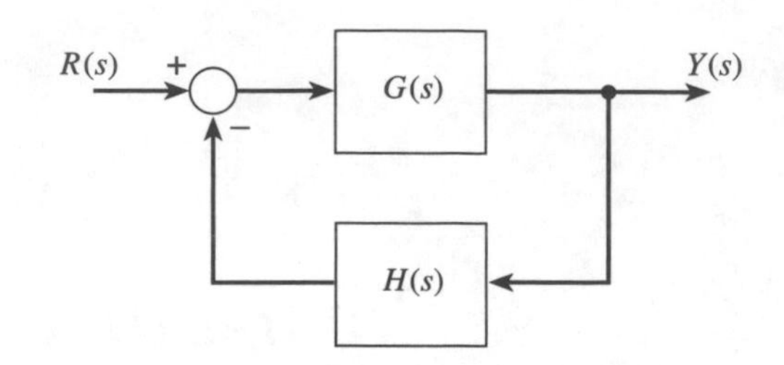

If the open-loop transfer function $GH(s)$ of an nth-order single-input/single-output system is available as the LTI object `GH` and contains a loop gain K, the root-locus plot can be generated by entering the command `rlocus(GH)`. The resulting plot will show how the closed-loop poles will move in the complex s-plane (where $s = \sigma + j\omega$) as the loop gain K varies from 0 to ∞, known as the 180°-locus. MATLAB will compute the n branches of the root locus and draw the plot, with real and imaginary axes that it selects. However, the numerical values of the roots corresponding to the values of K selected by MATLAB will not be available for viewing or printing. Alternatively, one can generate a set of gains, calculate the roots, and plot them with three separate commands. In either case, grid lines and a title must be added, and the user has the option of adjusting the plotting region by using MATLAB's `axis` command. If the command `axis equal` is given, the same scaling will be used for the real and the imaginary axes, thereby representing angles correctly and avoiding distortion of the locus. For example, a circle will appear as a circle.

Once the locus has been drawn, the `rlocfind` command can be used interactively to determine the value of K corresponding to any point on the locus that the user selects by clicking on it with the mouse. The values of all n closed-loop poles corresponding to this value of K can also be displayed by using the command in the form `[K,pCL] = rlocfind(GH)`.

The 0°-locus corresponds to either having a positive sign where the feedback path enters the summing junction in Figure 5.1 or having negative values for the loop gain K. To have MATLAB draw the 0°-locus, we can insert a negative sign in front of the numerator and proceed as before, as in the command `rlocus(-GH)`.

Now we will illustrate the construction of a root-locus plot with MATLAB and the determination of the gain corresponding to points on the locus.

EXAMPLE 5.1
Real Poles Only

Use MATLAB to draw the root locus for the feedback system whose open-loop transfer function is

$$GH(s) = \frac{K(s+5)}{(s+1)(s+3)(s+12)}$$

and comment on the applicability of the various root-locus properties. Find the value of K for which the two branches leave the real axis, and determine all the closed-loop poles for this value of K.

Solution The MATLAB commands in Script 5.1 will build the open-loop transfer function (with $K = 1$) in terms of the numerator and denominator polynomials and compute and plot the 180° locus, as shown in Figure 5.2.

FIGURE 5.2 *Root locus for Example 5.1 as produced by MATLAB*

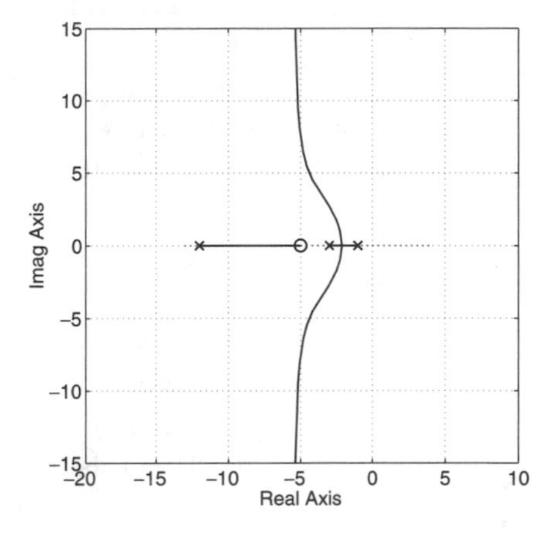

MATLAB Script

```
% Script 5.1:  Root locus for a system with real poles
zOL = -5                            % open-loop zero
pOL = [ -1; -3; -12 ]               % open-loop poles
G = zpk(zOL,pOL,1)                  % create G(s) as a ZP object
rlocus(G)                           % compute locus from the LTI object
axis equal                          % for uniform scaling on both axes
axis([-20 10 -15 15])               % show same area as in Fig. 5.2
set(findobj('marker','x'),'markersize',12) % larger 'X' for poles
set(findobj('marker','x'),'linewidth',1.5) % ....heavier line width
set(findobj('marker','o'),'markersize',9)  % larger 'O' for zeros
set(findobj('marker','o'),'linewidth',1.5) % ....heavier line width
%--  calc gain & other CL poles at point selected by user
[kk,clroots] = rlocfind(G)
```

We see that the real-axis portions are $-3 \le \sigma \le -1$ and $-12 \le \sigma \le -5$. There are three branches, with one starting from each of the open-loop poles at $s = -1, -3$, and -12. One branch terminates on the open-loop zero at $s = -5$. The other two branches go to infinity as $K \to \infty$, at angles of $\pm 90°$, and asymptotic to the vertical line through the point $\sigma_0 = [(-1 - 3 -$

$12) - (-5)]/(3 - 1) = -5.5$. Because all the open-loop poles and zeros are real, the angles of departure and arrival are either 0° or 180°.

Because there are two or more open-loop poles than zeros, the sum of the closed-loop poles remains constant. For $K = 0$ this sum is $-1 - 3 - 12 = -16$. As $K \to \infty$, this sum is $2(-5.5) - 5 = -16$, which agrees with the previous value.

The plot that MATLAB produces may include a larger region of the s-plane than is of interest, with correspondingly less detail where it is important. This is the case in Figure 5.2 where half of the area is devoted to the right half-plane, although the locus never leaves the left half-plane. Also, the default plots in MATLAB do not use uniform scaling in the horizontal and vertical directions, which will distort a root-locus plot. To correct for these conditions, the MATLAB `axis` command can be used in two ways. First, `axis equal` imposes equal scaling for both axes. Second, the command `axis([-20 10 -15 15])` will cause the plotting region to include the region between -20 and 10 on the horizontal (real) axis and between -15 and 15 on the vertical (imaginary) axis. Note the use of the commands `set` and `findobj` to increase both the sizes and the weights of the × and ○ symbols in order to make a more appealing plot.

Using the command `[kk,clroots]=rlocfind(G)` and placing the cursor at the point where the two complex branches leave the real axis gives $K \approx 3.4$ and the corresponding closed-loop poles are $s = -2.12, -2.12$, and -11.8. Figure 5.3 shows the root-locus plot with the pole locations indicated for several values of K.

WHAT IF? Redo the root-locus plot when the pole at $s = -12$ can be varied. Consider values in the interval $-40 \leq \sigma \leq -5$ and observe the effects on the locus. What happens to the large-gain asymptotes? What happens when the pole is at $s = -28$? ■

EXAMPLE 5.2
Real and Complex Poles

Use MATLAB to draw the root locus for

$$GH(s) = \frac{K(s + 8)}{s(s + 2)(s^2 + 8s + 32)}$$

and comment on the applicability of the various root-locus properties that are presented in most control systems textbooks. Calculate the angle at which the locus departs from the upper complex pole of $GH(s)$ and compare it with the computer-generated plot. Find the values of K for which (i) two branches cross into the right half-plane and (ii) the two branches leaving from the complex poles intersect at the real axis.

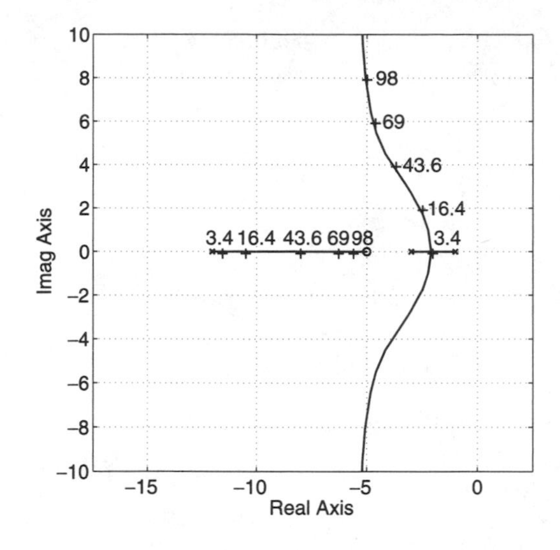

FIGURE 5.3 *Root locus for Example 5.1 with several gain values added and a different plotting region*

Solution The commands of MATLAB Script 5.2 will build the open-loop transfer function (with $K = 1$) in terms of the numerator and denominator polynomials and then compute and plot the root locus.

MATLAB Script

```
% Script 5.2: Root locus for a system with real & complex open-loop poles
den = conv([1 2 0],[1 8 32]) % denominator with 4 poles
G = tf([1 8],den)            % define numerator & build G(s)
Gzp = zpk(G)                 % display zero, poles, and gain of G(s)
rlocus(G)                    % compute locus
axis([-15 5 -10 10])         % adjust plotting area
[kk,clroots] = rlocfind(G)   % calculate gain value & poles
```

The plot produced by MATLAB is shown in Figure 5.4. We see that the real-axis portions are $\sigma \leq -8$ and $-2 \leq \sigma \leq 0$. There are four branches, with one starting from each of the open-loop poles at $s = 0, -2, -4 + j4$, and $-4 - j4$. One branch terminates on the open-loop zero at $s = -8$. The other three branches go to infinity as $K \to \infty$, at angles of $\pm 60°$ and $-180°$. The three large-gain asymptotes intersect at the point $\sigma_0 = [(0 - 2 - 4 - 4) - (-8)]/(4 - 1) = -2/3$.

The angle of departure from the upper complex pole can be calculated by hand (with a calculator). First we make the arbitrary designations $p_1 = -4 + j4$, $p_2 = -4 - j4$, $p_3 = 0$, $p_4 = -2$, and $z_1 = -8$. Then we write

$$\phi_1 = \arg(p_1 - z_1) - [\arg(p_1 - p_2) + \arg(p_1 - p_3) + \arg(p_1 - p_4)] + q180°$$

where q can be taken as any odd integer. Substituting the values of the poles and the zero, we obtain

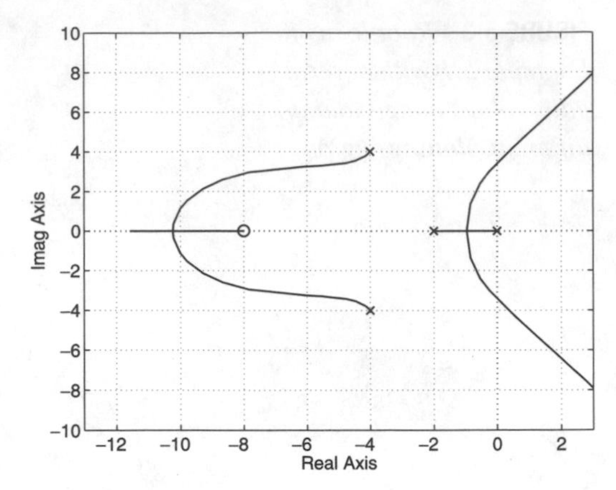

FIGURE 5.4 *Root locus for Example 5.2*

$$\begin{aligned}\phi_1 &= \arg(4 - j4) - [\arg(j8) + \arg(-4 + j4) + \arg(-2 + j4)] + 180 \\ &= 45 - (90 + 135 + 116.6) + 180 = -116.6^\circ\end{aligned}$$

The argument calculation can also be done in MATLAB using the function `angle`. An alternative to the calculations described above is to work from a plot or sketch of the poles and zeros, as shown in Figure 5.5, and write the equation for the angle criterion in terms of the unknown angles of the vectors from the other poles and the zero to p_1. Doing this and taking $q = 1$ give

$$45 - (\phi_1 + 90 + 135 + 116.6) = 180$$

which can be solved for ϕ_1.

Because the number of open-loop poles exceeds the number of open-loop zeros by at least two, the sum of the closed-loop poles remains constant as K varies. For $K = 0$ this sum is $0 - 2 - 4 - 4 = -10$. Hence the sum of the four closed-loop poles will be -10 for all values of K. Thus, when the two complex branches cross the imaginary axis ($\sigma = 0$), the other two closed-loop

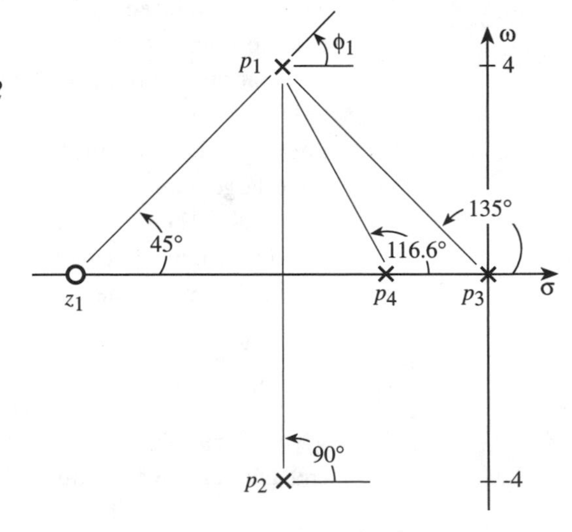

FIGURE 5.5 *Relationships used to determine the angle of departure in Example 5.2*

poles must have real parts of -5. From the locus in Figure 5.4 we see that they are at $s \approx -5 \pm j3.5$.

By using the `rlocfind` command we find that the two branches starting at the real poles cross into the right half-plane for $K \approx 45$ and that the two branches starting at the complex poles reach the real axis for $K \approx 2070$.

WHAT IF? Redo the locus of Example 5.2 when the location of the zero is varied over the interval $-20 \leq \sigma \leq -1$. Note what happens to the two branches that leave the complex poles. Also, compute the locus when the zero is at $s = -2.6$. ■

REINFORCEMENT PROBLEMS

Use MATLAB to draw root-locus plots for the open-loop transfer functions given in the following problems. In each case, draw an initial plot without specifying the boundaries of the plotting area. Then redo the plot using the `axis` command with appropriate arguments to best show the region of interest. After the plot has been drawn, answer any specific questions given in the problem statement. Also determine the value(s) of K that satisfies the stated condition(s) and give the corresponding closed-loop pole(s). Recall that σ_0 is the point at which the large-gain asymptotes intersect the real axis.

P5.1 Two real poles and one zero.

$$GH(s) = \frac{K(s+5)}{(s+1)(s+3)}$$

This is the system of Example 5.1 without the pole at $s = -12$. Because the number of poles exceeds the number of zeros by only 1, only one branch goes to infinity as $K \to \infty$. Use the `rlocfind` command to find the values of K for which the two branches intersect.

P5.2 Four poles.

$$GH(s) = \frac{K(s+5)}{(s+1)(s+3)(s+12)(s+20)}$$

This is the system of Example 5.1 with an additional pole at $s = -20$. Because $GH(s)$ has four poles and one zero, there are three branches that approach infinity as $K \to \infty$. After you plot the locus, calculate σ_0 and draw the large-gain asymptotes on the plot. Use `rlocfind` to determine K^*, the maximum gain for which the closed-loop system will be stable.

P5.3 Complex poles.

$$GH(s) = \frac{K(s+5)}{(s+1)(s+3-j2)(s+3+j2)}$$

This transfer function has a pair of complex poles in addition to a single real pole and a single real zero. Calculate σ_0 and the angle ϕ at which the locus departs from the pole at $s = -3 + j2$. Determine the value of K for which there is a closed-loop pole with an imaginary part of 10.

P5.4 Complex zeros.

$$GH(s) = \frac{K(s + 8 - j4)(s + 8 + j4)}{(s + 1)(s + 5)}$$

This transfer function has a pair of complex zeros and two real poles. Because the number of zeros equals the number of poles, no branches will approach infinity as K becomes large. Calculate the angle θ at which the locus arrives at the zero at $s = -8 + j4$. Find the values of K for which the two real branches meet and leave the real axis.

P5.5 Right half-plane pole.

$$GH(s) = \frac{K(s + 5)}{(s - 2)(s + 8)(s + 20)}$$

This transfer function has a pole in the right half of the s-plane, which means that the open-loop system is unstable. The closed-loop system will be unstable for low values of K because the branch that starts from the pole at $s = 2$ will be in the right half-plane. Once this branch has crossed into the left half-plane at $s = 0$, the closed-loop system will be stable because all three branches will be in the left half-plane. Determine this value of K and verify it by applying the magnitude criterion at $s = 0$.

P5.6 Double pole.

$$GH(s) = \frac{K(s + 5)(s + 10)}{s^2(s + 20)(s + 40)}$$

The locus has four branches, with two of them emanating from the double pole at $s = 0$. Because there are two zeros, there will be $4 - 2 = 2$ branches approaching infinity as $K \rightarrow \infty$. Determine the angles at which the branches leave $s = 0$ and find σ_0. Use `rlocfind` to determine the values of K for which the branches enter or leave the real axis.

P5.7 Triple pole.

$$GH(s) = \frac{K(s + 2)(s + 5)}{s^3}$$

This transfer function has a triple pole at $s = 0$ and two real zeros. There are three branches starting from the triple pole, with two of them starting into the right half-plane and them being pulled back into the left half-plane by the presence of the two zeros. One branch goes to infinity as $K \rightarrow \infty$, and the other two branches terminate on the two zeros.

After drawing the locus, determine the value of K^*, the gain for which the two complex branches cross into the left half-plane and give the corresponding value of ω. Also find the value of K for which the complex branches meet on the real axis and give the corresponding value of σ.

P5.8 Negative gain.

$$GH(s) = \frac{K(s+5)}{(s+1)(s+3)(s+12)(s+20)} \quad \text{where} \quad K < 0$$

To account for the fact that $K < 0$, we can insert a negative sign in front of the numerator and draw the locus for $K > 0$. Verify that the plot generated by MATLAB satisfies the properties of the 0°-locus. In particular, note that (i) the real-axis portions of the locus are those having an *even* number of poles and/or zeros to the right, and (ii) the angles of the three large-gain asymptotes are 0°, 120°, and –120°. Find the value of K for which the locus crosses into the right half-plane.

P5.9 State-space model. Make a root-locus plot of the state-space model (4.1) with

$$\mathbf{A} = \begin{bmatrix} 0 & 1 & 0 \\ -4 & -1 & 1 \\ 0 & 0 & -20 \end{bmatrix}, \quad \mathbf{B} = \begin{bmatrix} 0 \\ 0 \\ 20 \end{bmatrix}$$
$$\mathbf{C} = [1 \;\; 0 \;\; 0], \quad \mathbf{D} = [0]$$

using the MATLAB command `rlocus`. This root locus is the variation of the closed-loop system poles with the feedback control $u = Ky$ for $0 \leq K < \infty$.

VARIATION OF AN ARBITRARY PARAMETER

The procedures discussed in the previous sections are valid when the parameter to be varied is a gain in the feedback loop. However, it can be useful to construct the root-locus plot when the parameter being varied is not a gain in the loop. In such situations two options are available. One option is to rearrange the open-loop transfer function analytically so the parameter to be varied appears as if it were a loop gain. This approach can be used often but not always, and may require considerable work.

The other solution is to use MATLAB to calculate the *closed-loop* characteristic polynomial for specific values of the parameter and numerically solve for the roots of the polynomial. The roots can be collected into an array, with one row for each value of the parameter, and plotted in the s-plane.

We will illustrate this technique in the following example and present several problems that can be solved for practice.

EXAMPLE 5.3
Variable Zero Location

Write MATLAB code to compute and plot the roots of the closed-loop characteristic polynomial for the system whose open-loop transfer function is

$$GH(s) = \frac{5(s+\alpha)}{(s+1)(s+3)(s+12)}$$

where $2 \leq \alpha \leq 10$. Note that $GH(s)$ is the transfer function that was used in Example 5.1, with the gain fixed at $K = 5$ and with the zero variable from $s = -2$ to $s = -10$, rather than being fixed at $s = -5$.

Solution

A set of MATLAB commands that will compute the locus of closed-loop poles as α is varied from 2 to 10 is shown in Script 5.3. The solution is accomplished by constructing the closed-loop characteristic polynomial for a specific value of α and then determining the polynomial roots numerically. Once they have been computed, the roots are saved as a row in a matrix called `clpoles`. After the calculations have been done for the specified values of α, the closed-loop poles can be plotted in the s-plane as individual points, as shown in Figure 5.6. Smooth curves can be drawn through these points by hand if one desires, but MATLAB will not be able to do so because it does not know which points go with a specific branch.

MATLAB Script

```
% Script 5.3:  Root locus for variable open-loop zero location
temp = conv([1 1],[1 3]);         % denominator is product of 3 terms
den = conv(temp,[1 12])
K = 5                             % do rest of num inside the loop
clpoles = []; param = [];         % define arrays to hold results
for alpha = 2:10,                 % alpha varies from 2 to 10
  num = [0 0 K K*alpha]           % must be same length as den
  clpoly = num + den;             % form closed-loop characteristic poly
  clp = roots(clpoly);            % calculate closed-loop poles
  clpoles = [clpoles; clp'];      % save poles as nth row of array
  param = [param; alpha];         % save alpha as nth row of column vector
end
disp([param clpoles])             % print table of alpha and poles
plot(clpoles,'*')                 % plot points on the locus with * symbol
axis equal; axis([-4 0 -2 2]);    % replot area close to image axis
```

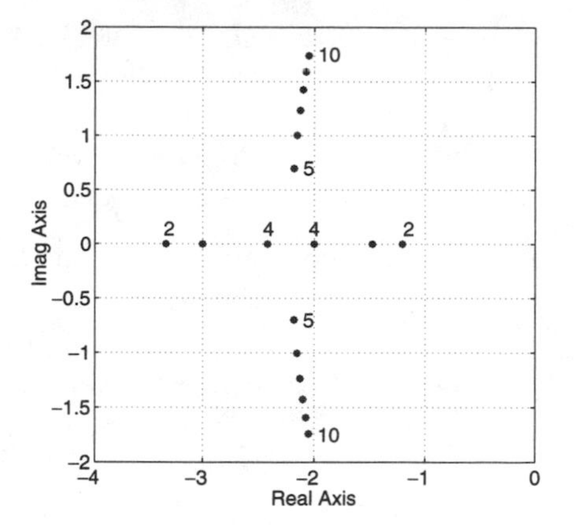

FIGURE 5.6 *Root-locus plot for Example 5.3, with a variable zero location*

The important point to recognize is that the closed-loop characteristic equation can be written as

$$\text{denominator}[GH(s)] + \text{numerator}[GH(s)] = 0 \tag{5.1}$$

Then the MATLAB command `roots` can be used to obtain the roots for that particular value of α. Doing the problem this way does not make use of any of the rules for constructing root-locus plots but does allow for very general expressions in the numerator and denominator polynomials. For this example the denominator does not depend on α and is formed in Script 5.3 outside the loop in which α is varied by using the `conv` command twice to multiply the three first-order polynomial terms. The numerator is formed inside the loop and is dependent on α. Note that the numerator polynomial `num` must be padded with two leading zeros so it is the same length as the denominator polynomial `den` for MATLAB to be able to add them.

Comment: The transfer functions used in Examples 5.1 and 5.3 are identical when the first example has $K = 5$ and the second example has $\alpha = 5$. The reader can verify that the closed-loop poles are the same (the values are: $s = -2.1806 \pm j0.6972$ and -11.64).

REINFORCEMENT PROBLEMS

Consider the open-loop transfer function

$$GH(s) = \frac{10(0.2s + 1)}{(\tau s + 1)(s^2 + 2\zeta\omega_n s + \omega_n^2)} \tag{5.2}$$

where ζ and ω_n are the damping ratio and undamped natural frequency of the complex open-loop poles, respectively, and τ is the time constant of the real pole. First, we let ω_n vary with ζ and τ fixed and plot the poles of the closed-loop system in the s-plane. Then we fix ζ and ω_n, let τ vary, and plot the new closed-loop poles. Finally, we fix ω_n and τ, and let ζ vary.

P5.10 Vary frequency ω_n. Using the open-loop transfer function (5.2) where $\zeta = 0.2$, $\tau = 4$, and $\omega_n = 1, 2, \ldots, 10$, write MATLAB code similar to that shown in Script 5.3 that will form the closed-loop characteristic polynomial for the specified parameter values, compute and tabulate the closed-loop poles, and plot them in the complex plane. You should find that two of the closed-loop poles are in the right half-plane when $\omega_n = 1$ and all three of the poles are in the left half-plane for $\omega_n = 2, 3, \ldots, 10$.

P5.11 Vary time constant τ. Repeat Problem 5.10 with $\zeta = 0.2$, $\omega_n = 5$, and $\tau = 0.5, 1.0, 1.5, \ldots, 8$. You should find that as τ is increased there are two complex closed-loop poles near $s = -1 \pm j5$ that move very little and one pole that moves from left to right along the real axis.

P5.12 Vary damping ratio ζ. Repeat the steps of Problems 5.10 and 5.11 with $\tau = 4$, $\omega_n = 5$, and the damping ratio $\zeta = 0, 0.1, 0.2, \ldots, 1.0$.

CROSS-CHECK The previous three problems deal with the same transfer function, under variation of three parameters. Each of the three root-locus plots has one set of roots that corresponds to the condition $\omega_n = 5, \tau = 4$, and $\zeta = 0.2$. Use MATLAB to form the closed-loop characteristic polynomial according to (5.1), and then make use of the `roots` command to show that the set of three closed-loop poles corresponding to this set of parameter values lies on each of the three root-locus plots. Give the values of the three poles. ■

EXPLORATION

E5.1 Effects on closed-loop poles. Exploratory Problem EP2.1 describes the manner in which the file `one_blk.m` can be used to perform a variety of analytical tasks on any single-block system whose transfer function exists in the workspace as the row vectors `num` and `den`. One of the analytical tools available on the menu of plots is the root-locus plot. Use this capability to explore the effects on the root locus (and thus on the closed-loop poles) of:

a. an open-loop zero in the right half-plane
b. an open-loop pole in the right half-plane
c. a pair of open-loop poles on the imaginary axis

d. a pole of multiplicity four at the origin of the s-plane (Problem 5.7 considered a triple pole)

E5.2 A useful root-locus property. It turns out that for any feedback system for which the number of open-loop poles exceeds the number of open-loop zeros by at least two, the sum of the closed-loop poles is independent of the gain K. Thus, if some roots move to the right as the gain increases, others must move to the left, such that the sum of the roots does not vary. In fact, the sum of the roots must equal the sum of the open-loop poles. Why? Construct one or more root-locus plots that satisfy this criterion, and use the `rlocfind` command to determine the roots for several values of the loop gain. Does the sum remain constant? Is the sum equal to the sum of the open-loop poles? Note that the open-loop transfer functions given in Problems 5.2, 5.3, 5.5, 5.6, and 5.8 all satisfy this condition.

COMPREHENSIVE PROBLEMS

CP5.1 Electric power generation system. The voltage feedback system with a lag controller for the electric power system in Comprehensive Problem CP3.1 was shown to become unstable when the controller gain K exceeds about 50. Make a root-locus plot of this voltage-feedback system by varying K, with separate expanded plots about the origin and the electromechanical mode for examining the detail of the root locus. You will find that the root loci originating from the lightly damped poles due to the electromechanical mode will migrate into the right half-plane and then move back into the left half-plane, terminating at the neighboring complex zeros. Use `rlocfind` to find the values of K at the imaginary-axis crossings.

CP5.2 Satellite with reference input. Figure 3.8 shows the block-diagram representation of the satellite with reaction wheel when the input is the desired pointing angle θ_{des}. Use the MATLAB file `sat.m` to obtain the open-loop transfer function from the reaction wheel motor torque τ_m to the satellite's pointing angle θ, as described in Appendix A. Then construct M-files for solving the problems described below. For each of the controllers discussed, select one or more values of the controller gain based on the root-locus and the use of the command `rlocfind` and plot the response of the satellite to a step change in the desired pointing angle. Try to relate what the root-locus plot shows to what you see in the step response.

a. It was stated in Comprehensive Problem CP 3.2 that in order for the closed-loop satellite to be stable, the gain of the controller must be *negative*. Use root-locus plots to verify that this is true by assuming a proportional-only controller and drawing both the 180° and 0° loci. As suggested in the statement of Problem 5.8, you can obtain the 0°-locus by using the negative of the numerator in the `rlocus` command, as in `rlocus (-G)`. You should find that the 180°-locus always has a branch in the right half-plane, whereas the 0°-locus remains in the left half-plane for all $-\infty < K < 0$.

b. Put a lead controller (3.3) with its gain set to -1 in series with the open-loop satellite model (with input τ_m and output θ) and draw the root-locus plot. To get started, place the lead zero at $s = -1$ and use $\alpha = 20$. Use the `axis` command to enlarge the area close to the origin so you can see the details of the loci. You should find that the closed-loop system has two poles in the LHP for all negative controller gains. Also there will be a pole-zero pair right at the origin. Physically this corresponds to the fact that the controlled satellite can exhibit a steady-state pointing error if the speed of the reaction wheel is not zero, because of viscous friction between the wheel and the satellite.

c. Construct a lead-lag controller (3.4) by adding a lag element to the lead controller, where the lag zero is 1/10 of the lead zero. After redrawing the plot, you should find that the root locus is affected close to the origin of the s-plane but is virtually the same as for the lead controller elsewhere. In order to see the full details of the root locus, use the `axis` command to draw a series of four plots that have vertical and horizontal dimensions of approximately 40, 4.0, 1.0, and 0.1 units in the s-plane. You should find that for very small gains two complex branches move into the right half-plane and then cross back into the left half-plane as the gain increases. This says that with a lead-lag controller we should expect the closed-loop system to be *conditionally* stable because the feedback system could go unstable if the gain falls below a minimum value.

CP5.3 Stick balancing system.

a. Using the transfer function from the motor voltage u to the stick angle θ for the rigid stick case, draw a root-locus plot that shows the locations of the closed-loop poles as the gain ($K_c > 0$) of a proportional controller is varied, where θ is the variable being fed back. Use the resulting locus to show that it is not possible to stabilize the rigid-stick system by feeding back the stick angle alone. To be sure that the problem is not just the result of having the wrong sign for the control gain, draw the 0°-locus ($K_c < 0$) and show that the same conclusion follows. It will turn out that we will need at least the cart position x or the cart velocity v in addition to θ to construct a control system that is capable of balancing the stick.

b. Repeat the steps of part (a) but using the transfer function from the motor voltage u to the cart position x. For the 180°-locus it may appear that all of the roots will be in the left half-plane, but this will not be the case. What appears to be a pole-zero cancellation on the positive real axis at $s = +2.12$

is actually a very short branch of the locus starting at the open-loop pole at $s = +2.8233$ and ending at the open-loop zero at $s = +2.8218$. Actually, the closed-loop system would be unstable even if the pole and zero cancelled exactly. Such a cancellation would happen only if the motion of the stick had no effect on the position of the cart, so the stick angle was *unobservable* from measurements of the cart position. But the stick would still be there and would fall over regardless of what the cart did because of its unstable equilibrium (which gives us the pole in the right half-plane).

CP5.4 Hydro-turbine system. In each part of this problem, you are to construct a feedback system for controlling the mechanical power produced by the turbine according to the block diagram in Figure A.8 in Appendix A. The transfer function from the actuator input to the power developed can be obtained in TF form by running the MATLAB file `hydro.m`.

a. Draw the root locus for the hydro-turbine plant with a proportional controller. You should find that there are two open-loop poles on the negative real axis and a single zero on the positive real axis. The two branches of the root locus will start in the left half-plane for very low gains, but both branches will cross the imaginary axis and remain in the right half-plane as the gain increases. Use `rlocfind` to determine the values of the gain K_p^* and the frequency $\omega*$ at which the loci cross into the right half-plane.
b. Replace the proportional controller with a lead controller (3.3) having a pole/zero ratio of $\alpha = 20$. Experiment with the location of the controller's zero to see what effect its location has on the root loci. You will have to use the `axis` command to specify a region close to the origin to observe the details of the plot. Try selecting the zero to cancel one of the plant's poles that are at $s = -1$ and -0.5. Use `rlocfind` to determine the gain required to make a specific point on the locus be a closed-loop pole. Simulate the step response of the closed-loop system for several choices of the gain K_{lead} and the zero location z_{lead}.
c. Use a lead-lag controller (3.4) with the same values for the lead zero and α as above, and with the lag zero set to $1/10$ of the lead zero. Account for each of the open-loop poles and zeros; there should be three zeros and four poles. The locus with the lead-lag controller should not look much different from that with the lead controller except for a short branch connecting the lag pole and the lag zero, which is on the real axis just to the left of the origin. Although this pole-zero pair, often referred to as a dipole, does not have much effect on the root-locus plot, it will increase the low-frequency gain of the open-loop system by the factor α and will reduce the steady-state error to a step input by the same factor. You can see this beneficial effect by comparing the step responses of the closed-loop system with and without the lag component present.

SUMMARY

We have shown how commands of the Control System Toolbox can be used to draw root-locus plots and to determine the value of the loop gain that will cause a user-selected point on any branch of the locus to be a pole of the closed-loop transfer function. We have used the root locus to illustrate the effects on a system's closed-loop poles of open-loop zeros and poles in the right half-plane, and lightly damped modes. We have also shown how a root-locus plot can be generated for a parameter that is not a loop gain, such as a time constant, damping ratio, or undamped natural frequency associated with the system's open-loop transfer function.

MATLAB FUNCTIONS USED

Function	*Purpose and Use*	*Toolbox*
+	Given two LTI objects, the + operator forms their parallel connection.	Control System
angle	Given a complex number, **angle** returns the phase angle, in radians.	MATLAB
axis	**axis([xmin xmax ymin ymax])** specifies the plotting area. **axis equal** forces uniform scaling for the real and imaginary axes.	MATLAB
conv	Given two row vectors containing the coefficients of two polynomials, **conv** returns a row vector containing the coefficients of the product of the two polynomials.	MATLAB
findobj	Given a set of handle-graphics objects, **findobj** returns the handles of those objects having the specified property.	MATLAB
rlocfind	Given a TF or state-space description of an open-loop system, **rlocfind** allows the user to select any point on the locus with the mouse and returns the value of the loop gain that will make that point be a closed-loop pole. It also returns the values of all the closed-loop poles for that gain value.	Control System

rlocus	Given a TF or state-space description of an open-loop system, **rlocus** produces a root-locus plot that shows the locations of the closed-loop poles in the s-plane as the loop gain varies from 0 to infinity.	Control System
roots	Given a row vector containing the coefficients of a polynomial $P(s)$, **roots** returns the solutions of $P(s) = 0$.	MATLAB
set	Given a handle-graphics object, **set** assigns the value of the specified property.	Control System
tf	Given numerator and denominator polynomials, **tf** creates the system model as a TF object. The command also converts zero-pole-gain or state-space models to TF form.	Control System
zpk	Given a system's zeros, poles, and gain, **zpk** creates the system model as a ZPK object. The command also converts transfer-function or state-space models to ZPK form.	Control System

ANSWERS

P5.1 Branches intersect for $K \approx 0.34$ and 11.7

P5.2 $\sigma_0 = -10.33$ and $K^* \approx 6540$

P5.3 $\sigma_0 = -1.0$, $\phi = 0°$, and $K \approx 93$

P5.4 $\theta = 7°$ and branches leave real axis when $K \approx 0.10$

P5.5 Minimum gain for stability is $K = 64$

P5.6 $\phi = \pm 90°$; $\sigma_0 = -20$; branches enter or leave real axis for $K = 0$, 170, and 4394

P5.7 a. Branches enter left half-plane for $K \approx 1.58$ and frequency is $\omega = 3.30$
b. Branches meet on real axis for $K \approx 24.6$ and $\sigma_0 = -11.36$

P5.8 Locus enters right half-plane for $K \approx -170$

CHAPTER 6

Frequency-Response Analysis

PREVIEW

In addition to root-locus analysis, frequency-response analysis also provides a control designer with important insights of a linear, time-invariant system. Frequency response is useful for determining the steady-state response to sinusoidal input signals and noise. Furthermore, it allows the determination of the stability of a feedback system in the frequency domain, establishing the gain- and phase-margin concept. Based on these margins, control design methods aimed at modifying the frequency response of the compensated system to achieve improved margins have been developed. In this chapter, we illustrate with MATLAB the computation of the sinusoidal steady-state response and the plotting of the frequency response for a system given in either the transfer-function or the state-space form. Then frequency-domain techniques to determine feedback system stability will be illustrated. The design of controllers using frequency response will be discussed in Chapter 9.

SINUSOIDAL STEADY-STATE RESPONSE

The frequency response of a linear, time-invariant system $G(s)$ at a frequency ω rad/s is defined as

$$G(j\omega) = G(s)|_{s=j\omega} = |G(j\omega)| \arg[G(j\omega)] \tag{6.1}$$

where $|G(j\omega)|$ denotes the gain and $\arg[G(j\omega)]$ the phase angle of $G(j\omega)$. In terms of a system's state-space representation, its frequency response is given by

$$G(j\omega) = \mathbf{C}(j\omega\mathbf{I} - \mathbf{A})^{-1}\mathbf{B} + \mathbf{D} \tag{6.2}$$

where $\mathbf{I}$ is an identity matrix of the same dimension as $\mathbf{A}$.

The frequency response of a system can be computed using the Control System Toolbox function `bode`. If $G(s)$ exists as a LTI object, in TF, ZPK, or SS form, the frequency response $G(j\omega)$ can be obtained using the command `[mag,phase] = bode(G,omega)`, where the argument `omega` is a column vector of frequency points in rad/s. The output variable `mag` is a column vector of the magnitude of the frequency response, and `phase` is a column vector of the corresponding phase angles in degrees.

If the frequency response of $G(s)$ at a specific frequency ω_0 is to be computed, then `omega` is the scalar ω_0. This application is relevant in computing the sinusoidal steady-state response. If the input to $G(s)$ is a sinusoidal signal $u(t) = a\cos(\omega_0 t + \theta)$ and $G(s)$ is asymptotically stable, then the sinusoidal steady-state response of $G(s)$, after all the transients have decayed, is given by

$$y_{ss}(t) = a|G(j\omega_0)| \cos(\omega_0 t + \theta + \arg[G(j\omega_0)]) \tag{6.3}$$

where $|G(j\omega_0)|$ acts as an amplification factor, and $\arg[G(j\omega_0)]$ is the phase shift. The following example illustrates these properties.

EXAMPLE 6.1
Sinusoidal Steady-State Response

Simulate the complete response of the system

$$G(s) = \frac{10s + 50}{s^2 + 4s + 3}$$

to the sinusoidal input signal $u(t) = 2\cos(5t + 30°)$ over the interval $0 \le t \le 6$ s, assuming zero initial conditions. Then find the frequency response $G(j\omega)$ for $\omega = 5$ rad/s, and compute $y_{ss}(t)$ from (6.3). Draw $u(t)$, $y(t)$ and $y_{ss}(t)$ in a single plot, and comment on their relationships to one another.

Solution

Because the poles of $G(s)$ are at $s = -1$ and -3, the system is asymptotically stable, and the sinusoidal steady-state response can be computed using its frequency response. The MATLAB commands that implement the required calculations are given in Script 6.1. The input signal is generated by `u = 2*cos(5*t + 30*pi/180)`, and the `lsim` command simulates the complete output response, where the time `t = [0:0.06:6]'` is a column vector. The frequency response $G(j5)$ is computed using the `bode` command, yielding $|G(j5)| = 2.378$ and $\arg[G(j5)] = -92.73°$. Then $y_{ss}(t)$ is generated from (6.3). The plots of $u(t)$, $y_{ss}(t)$, and $y(t)$ produced by MATLAB are shown in Figure 6.1. Note that after about 4 seconds, $y(t)$ is almost identical to $y_{ss}(t)$.

MATLAB Script

```
% Script 6.1: Sinusoidal steady-state response computation
G = tf([ 10 50 ], [ 1 4 3 ])                 % create G(s) as a TF object
t = [0:0.06:6]';                             % time in a column vector
u = 2*cos(5*t + 30*pi/180);                  % input signal
y = lsim(G,u,t);                             % complete response, no IC's
[mag,phase] = bode(G,5)                      % frequency response at 5 rad/s
yss = 2*mag*cos(5*t+(30+phase)*pi/180);      % steady-state response per (6.3)
plot(t,u,'-',t,y,'--',t,yss,'-.')            % plot time responses
```


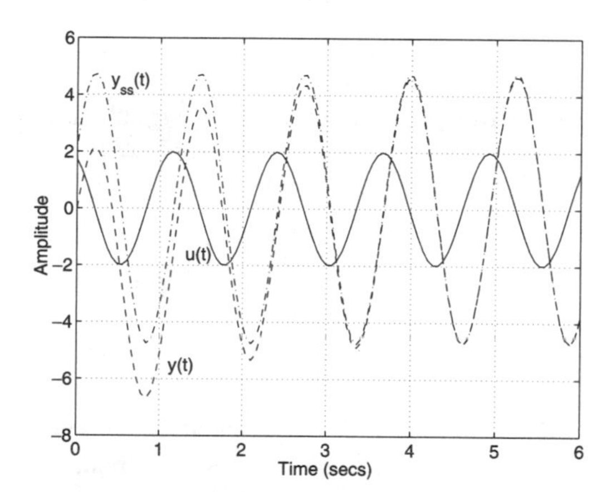

FIGURE 6.1 *Input $u(t)$, complete response $y(t)$, and steady-state response $y_{ss}(t)$ for Example 6.1*

Comment: The frequency response $G(j5)$ can be found from Figure 6.1 by comparing $y(t)$ to $u(t)$ after the initial system transients have decayed. The magnitude $|G(j5)|$ is the ratio of the amplitude of $y(t)$ to that of $u(t)$. The phase angle $\arg[G(j5)]$ is $(t_2 - t_1)/T \times 360°$, where t_1 is the time instant corresponding to a zero crossing of $u(t)$ from negative to positive values, t_2 is the time instant corresponding to the first zero crossing of $y(t)$ from negative to positive values after t_1, and $T = 2\pi/5$ is the period of one oscillation. In fact, the frequency response of a system can be found experimentally by repeatedly exciting the system with single-frequency sinusoidal input signals whose frequencies are in the range of interest. For each input signal, the initial transients are allowed to decay and the sinusoidal steady-state response is measured to perform the frequency-response calculation.

WHAT IF? Repeat Example 6.1 using the input signals $u(t) = 2\cos(t + 30°)$ and $u(t) = 2\cos(20t + 30°)$. Since the frequency response $G(j\omega)$ is a function of ω, the resulting output response will have different amplifications and phase shifts. ■

REINFORCEMENT PROBLEMS

In each of the following problems, simulate the output $y(t)$ assuming zero initial conditions for the given system $G(s)$ and the sinusoidal input signal $u(t)$. Find the frequency response $G(j\omega)$ at the input signal frequency, and use (6.3) to compute $y_{ss}(t)$. Plot and compare $u(t)$, $y(t)$, and $y_{ss}(t)$.

P6.1 Single-frequency input signal.

$$G(s) = \frac{2s^2 + 3s + 4}{s^3 + 2s^2 + 2.5s + 0.5} \qquad u(t) = 4\cos(2t - 60°)$$

P6.2 Multiple-frequency input signal, low-pass filter.

$$G(s) = \frac{0.1s + 1}{s + 1} \qquad u(t) = \sin 0.1t + \cos(t + 45°) - \sin(10t - 30°)$$

The input signal $u(t)$ consists of three sinusoidal components. Since the system is linear, the overall sinusoidal response is obtained by adding the individual sinusoidal responses. Find the steady-state response of the individual sinusoidal components by using the column vector `omega = [ 0.1 1 10 ]'` in the function `bode`. Note that although the individual input sinusoidal components have identical amplitudes, the amplitude of the 10 rad/s component in the output signal is about an order of magnitude smaller than that of the 0.1 rad/s component.

P6.3 Multiple-frequency input signal, high-pass filter.

$$G(s) = \frac{s + 1}{s + 10} \qquad u(t) = \sin 0.1t + \cos(t + 45°) - \sin(10t - 30°)$$

P6.4 DC gain.

$$G(s) = \frac{2s^2 + 3s + 4}{s^3 + 2s^2 + 2.5s + 0.5} \qquad u(t) = \text{unit step}$$

Constant input signals can be treated as sinusoidal signals with $\omega = 0$ in the function `bode`. Alternatively, the function `dcgain` can also be used.

P6.5 State-space model.

$$\mathbf{A} = \begin{bmatrix} 0 & 1 & 0 \\ -4 & -1 & 1 \\ 0 & 0 & -20 \end{bmatrix}, \qquad \mathbf{B} = \begin{bmatrix} 0 \\ 0 \\ 20 \end{bmatrix}$$

$$\mathbf{C} = [1 \quad 0 \quad 0], \qquad \mathbf{D} = [0]$$

and the input

$$u(t) = 4 \cos 2t$$

FREQUENCY-RESPONSE PLOTS

Control system analysis and design typically require an examination of the frequency response of a system $G(s)$ over a range of frequencies of interest. By specifying the frequencies ω in the column vector `omega`, we can use MATLAB to compute the frequency response $G(j\omega)$ and plot the results. Given a system representation as a LTI object `G` in any of the three forms that we have been considering (TF, ZPK, or SS), the Control System Toolbox provides functions to generate three common graphical forms of presenting the frequency response:

1. Bode plot—the command `bode(G,omega)` generates a semi-logarithmic plot of the magnitude $|G(j\omega)|$ in decibels[1] versus ω and a separate semi-logarithmic plot of the phase angle $\arg[G(j\omega)]$ in degrees versus ω.
2. Nichols plot—the command `nichols(G,omega)` generates a plot of the magnitude $|G(j\omega)|$ in decibels versus the phase angle $\arg[G(j\omega)]$ in degrees.
3. Nyquist plot—the command `nyquist(G,omega)` generates a $G(j\omega)$ plot in a complex plane where the axes are $\mathrm{Re}[G(j\omega)]$ and $\mathrm{Im}[G(j\omega)]$, the real and imaginary parts of $G(j\omega)$.

When the functions `bode`, `nichols`, and `nyquist` are used with no output variables, frequency-response plots are automatically generated. If output variables are specified, they will contain the frequency response in numerical form, but the user must enter separate plotting commands. The command `[mag,phase] = bode(G,omega)` returns the magnitude in the column vector `mag` and the phase angles in degrees in the column vector `phase`. The magnitude can be converted to decibels using the command `mag_dB = 20*log10(mag)`. Similarly, using the command

[1]Whenever a magnitude value is expressed in decibels, it will be labeled as such.

[re,im] = nyquist(G,omega), the output column vectors re and im contain the real and imaginary parts, respectively, of the frequency response.

The variable omega is the vector of frequency points in rad/s. Typically we use the MATLAB function logspace to generate frequency values that are uniformly spaced on the logarithmic scale. If omega is omitted as an input variable, then these plotting functions will use a frequency vector computed by MATLAB based on the poles and zeros of the system. The user can obtain omega by listing it as the third output variable.

The commands grid and ngrid can be called to add grid lines to the Bode and Nichols plots, respectively.

EXAMPLE 6.2
Bode, Nichols, and Nyquist Plots

Draw the Bode, Nichols, and Nyquist plots for the transfer function

$$G(s) = \frac{1280s + 640}{s^4 + 24.2s^3 + 1604.81s^2 + 320.24s + 16}$$

Solution

We use the commands bode, nichols, and nyquist to plot the frequency response

$$G(j\omega) = \frac{1280j\omega + 640}{(j\omega)^4 + 24.2(j\omega)^3 + 1604.81(j\omega)^2 + 320.24j\omega + 16}$$

Script 6.2 will compute the frequency response and do the plotting. Although we can let MATLAB generate its own frequency vector for making the Bode plot, we choose to illustrate here the use of logspace for that purpose. The command w = logspace(-2,3,100) will generate a row vector w containing 100 points from 10^{-2} to 10^3 that are uniformly spaced on the logarithmic scale. The resulting Bode plot, generated directly by the bode command, is shown in Figure 6.2(a). The grid lines shown in the plot are added automatically by the function bode.

MATLAB Script

```
% Script 6.2: Bode, Nichols, and Nyquist plots
numG = [1280 640],            % creat G(s) as TF object
denG = [1 24.2 1604.81 320.24 16]
G = tf(numG,denG)
w = logspace(-2,3,100)'   % logarithmically-spaced points as column
bode(G,w)                 % Bode plot
nichols(G,w)              % Nichols plot
axis([-270 0 -40 40])     % adjust plot area
grid;ngrid                % add rectangular & Nichols grid lines
nyquist(G)                % Nyquist plot
axis equal                % adjust aspect ratio so circle looks like circle
```

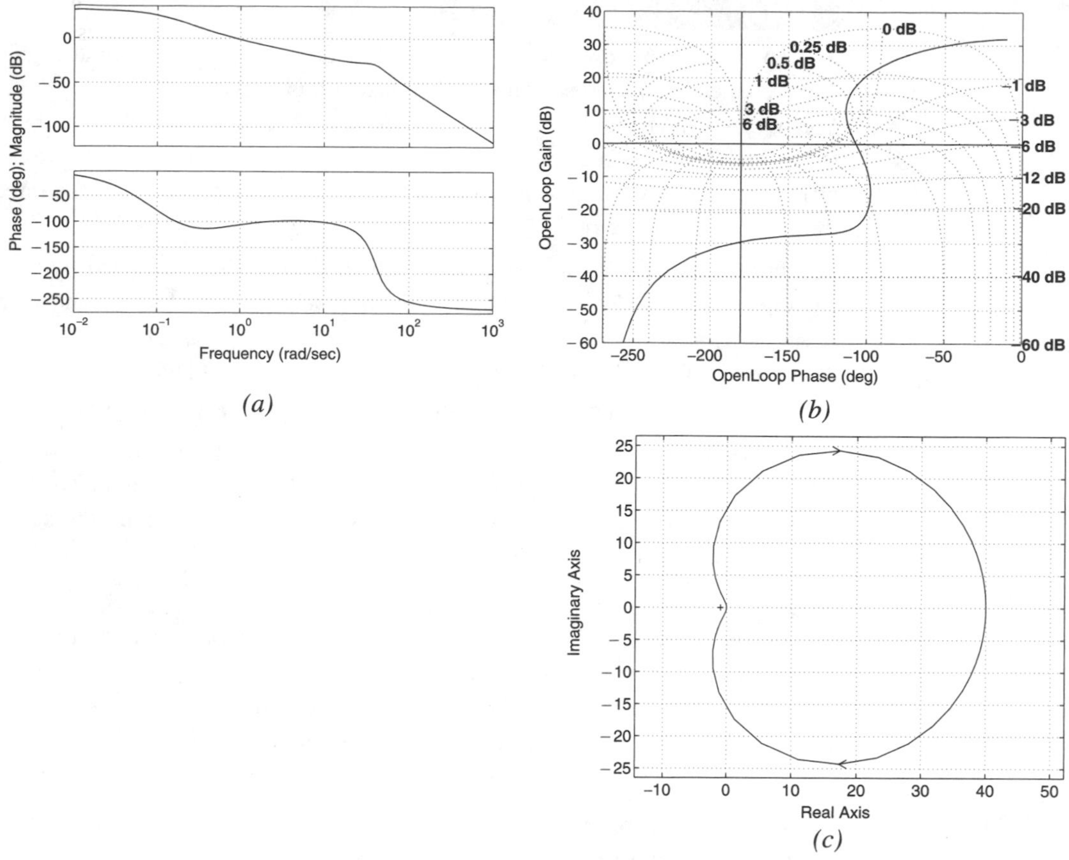

FIGURE 6.2 *Frequency response plots for Example 6.2 (a) Bode plot (b) Nichols plot (c) Nyquist plot*

The same frequency vector `w` is used in the `nichols` command to generate the Nichols plot. We then use the `axis` command to adjust the plot area to provide a better view of the plot and the `grid` command to add the grid lines. Finally, we use the `ngrid` command to add Nichols grid lines to the plot, which is shown in Figure 6.2(b). These grid lines, ranging from -60 dB to 6 dB, allow us to find the magnitude of the frequency response of the closed-loop transfer function.

For the Nyquist plot, we let the function `nyquist` compute the desired frequency range to be used, by omitting the third argument. The `axis equal` command is used to preserve the angles in the plot, resulting in Figure 6.2(c). In the Nyquist plot, the lower half corresponds to the frequency response for $\omega > 0$, while the upper half corresponds to that for $\omega < 0$. The function `nyquist` also marks the encirclement direction of the plot, and puts a "+" at the critical point $(0, -1)$ for making encirclement counts (see next section).

WHAT IF?

a. Repeat Example 6.2 for the transfer function $KG(s)$ in which the gain K is connected in series with $G(s)$. Make plots for $K = 0.1$, 2, and 10. You will see that the magnitude of the frequency response is changed by a factor of K over all frequencies, while the phase remains unchanged. Note the different ways in which this magnitude change is reflected in each of the three frequency-response plots.

b. Repeat Example 6.2 for the transfer function $G_c(s)G(s)$ obtained by connecting $G_c(s) = (0.2s + 1)/(2s + 1)$, the transfer function of a lag controller, in series with $G(s)$. Note that the frequency response $G_c(j\omega)G(j\omega)$ is simply the product of the individual frequency responses $G_c(j\omega)$ and $G(j\omega)$.

c. Repeat Example 6.2 by changing the numerator of $G(s)$ to $-1280s + 640$ such that $G(s)$ is non-minimum phase. Find the poles and zero of the resulting $G(s)$, and compare them to the poles and zero of the original system. You will observe that the magnitude of the frequency response will remain the same, but the phase will be different. ■

Of the three different forms of frequency-response plots, an approximate Bode plot can be sketched manually by using asymptotes if the *corner frequencies* are known. The following example shows how the asymptotes can be determined from a knowledge of the system's poles and zeros.

EXAMPLE 6.3 *Corner Frequencies and Asymptotes*

Identify the corner frequencies, and use MATLAB to draw the magnitude asymptotes for the Bode plot of Example 6.2.

Solution

We begin by creating `G` as a TF object and using the `zpkdata` command to extract its zeroes and poles as the MATLAB variables `z` and `p`. We find that the zero of the system $G(s)$ is at $s = -0.5$ and the poles are at $s = -0.1, -0.1$, and $-12 \pm j38.16$.

Then, the Control System Toolbox command `damp` is used to determine the damping ratios and natural frequencies of the zeros and poles of $G(s)$. With this information, we can express $G(s)$ in a special zero-pole form as

$$G(s) = \frac{40(1 + s/0.5)}{(1 + s/0.1)^2(1 + 2 \times 0.3s/40 + s^2/40^2)} \tag{6.4}$$

where the complex poles are represented by the quadratic form in the denominator with the natural frequency $\omega_n = 40$ and the damping ratio $\zeta = 0.3$. Equation (6.4) is in the *Bode* form where the constant term in each factor is unity. The corner frequencies of $G(j\omega)$ are at $\omega = 0.1, 0.5$, and 40 rad/s, since the magnitude asymptotes change slope at those frequencies.

As indicated in MATLAB Script 6.3, we can superpose the magnitude asymptotes based on these corner frequencies on the plot of magnitude versus frequency. First we use bode to generate the frequency response and plot the magnitude denoted by the dashed curve in Figure 6.3. We use the hold on command to add the asymptotes to this plot. The low frequency asymptote of 32 dB in cf1_dB is obtained by using the function dcgain and converting the result to decibels. This value is plotted as a straight line from a low frequency point to the first corner frequency $\omega = 0.1$ rad/s. Between the corner frequencies 0.1 and 0.5 rad/s, the magnitude asymptote decreases at a slope of -40 dB/decade, because $G(s)$ has a double pole at $s = -0.1$. The value of the asymptote at $\omega = 0.5$ rad/s is computed in cf2_dB. Then a straight line connecting the asymptote values at $\omega = 0.1$ and 0.5 rad/s is plotted. This line has a slope of -40 dB/decade. The remaining magnitude asymptotes can be generated in a similar manner, as shown in Script 6.3. At $\omega = 0.5$ rad/s, the zero at $s = -0.5$ changes the slope of the magnitude asymptote to -20 dB/decade. At $\omega = 40$ rad/s, the complex poles change the slope of the magnitude asymptote to -60 dB/decade. These asymptotes, with their slopes labeled, are shown as the solid curve in Figure 6.3.

MATLAB Script

```
% Script 6.3:  plot Bode magnitude asymptotes
numG = [1280 640]                          % create G(s) as TF object
denG = [1 24.2 1604.81 320.24 16]
G = tf(numG,denG)
[z,p,k] = zpkdata(G,'v')
% compute system poles and zeros
damp(z),damp(p)                 % determine damping ratios & natural freqs.
abs(z),abs(p)                   % break frequencies due to zeros & poles
w = logspace(-2,3,100)';                   % logarithmically-spaced points
[mag,phase] = bode(G,w)                    % compute frequency response
% reshape a 1x1x100 matrix into a 100x1 matrix
mag = reshape(mag,[100 1]);
semilogx(w,20*log10(mag),'--')             % make magnitude plot
hold on
lfg = dcgain(G)                            % low frequency gain
cf1_dB = 20*log10(lfg)                     % mag of low frequency asymp
plot([0.01;0.1],[cf1_dB;cf1_dB],'-')       % straight line plot
cf2_dB = cf1_dB - 40*log10(0.5/0.1)        % mag at w = 0.5 rad/s
plot([0.1;0.5],[cf1_dB;cf2_dB],'-')        % -40 dB/dec slope
cf3_dB = cf2_dB - 20*log10(40/0.5)         % mag at w = 40 rad/s
plot([0.5;40],[cf2_dB;cf3_dB],'-')         % -20 dB/dec slope
end_dB = cf3_dB - 60*log10(1000/40)        % mag at w = 1000 rad/s
plot([40;1000],[cf3_dB;end_dB],'-')        % -60 dB/dec slope
```

Continues

```
hold off                          % release plot
text(0.23,25,'-40 db/dec')        % add labels for asymptotes
text(1.5,-25,'-20 db/dec')
text(15,-70,'-60 db/dec')
axis([0.01 1000 -100 40])         % readjust y axis
```

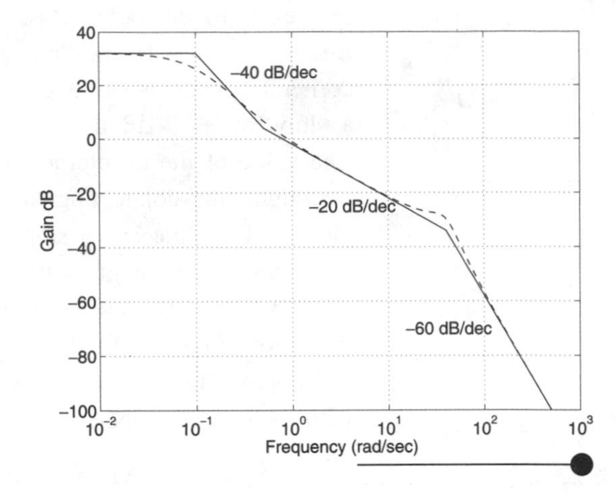

FIGURE 6.3 *Bode magnitude plot with asymptotes added for Example 6.3*

REINFORCEMENT PROBLEMS

In each of the following problems, draw the frequency-response plots for the given transfer function $G(s)$. Draw a separate Bode magnitude plot and identify the corner frequencies, and draw the magnitude asymptotes on the same plot. Label the slopes of the asymptotes.

P6.6 Complex poles, different damping ratios. Draw the Bode plots for

$$G(s) = \frac{1}{s^2 + 2\zeta s + 1}$$

where the damping ratio $\zeta = 0.1$, 0.4, 0.7, and 1.0, on a single plot. Repeat for the Nichols and Nyquist plots. For small ζ, note the peaks in the Bode magnitude plots. In the Nyquist plots, these peaks cause the frequency response to go outside the unit circle centered at the origin of the s-plane; that is, the magnitude is greater than unity.

P6.7 Lag controller. Draw the Bode plots for

$$G_c(s) = \frac{s + 1}{s + 1/\alpha}$$

where $\alpha = 2$, 5, and 20. Note the low-pass and phase-lag (negative phase) characteristics of the frequency response. For each α, identify the frequency at which the phase is minimum.

P6.8 Lead controller. Draw the Bode plots for

$$G_c(s) = \frac{s + 1}{s + \alpha}$$

where $\alpha = 2$, 5, and 20. Note the high-pass and phase-lead (positive phase) characteristics of the frequency response. For each α, identify the frequency at which the phase is maximum.

P6.9 System with real poles. Draw the Bode, Nichols, and Nyquist plots for

$$G(s) = \frac{1000(s + 5)}{(s + 1)(s + 3)(s + 12)(s + 20)}$$

This is the system in Problem 5.2 with $K = 1000$.

P6.10 Notch filter. Draw the Bode, Nichols, and Nyquist plots for

$$G(s) = \frac{40(s^2 + s + 32)}{(s + 40)(s^2 + 4s + 32)}$$

Explain why a system having such a transfer function is referred to as a "notch filter."

P6.11 State-space model. Draw the Bode, Nichols, and Nyquist plots for the state-space model of Problem 6.5.

FREQUENCY-DOMAIN STABILITY ANALYSIS

Based on the Nyquist criterion, frequency-response information can be used to determine the stability of the feedback system in Figure 6.4. We will denote the open-loop system transfer function $G(s)H(s)$ by $GH(s)$. The Nyquist criterion states that the number of unstable poles of the closed-loop system is equal to the number of clockwise encirclements about the critical point $(-1, 0)$ by the Nyquist plot minus the number of open-loop unstable poles of $GH(s)$. If $GH(s)$ is stable, the Nyquist criterion reduces to the stability condition that $k_m > 1$ and $\phi_m > 0$, where k_m is the *gain margin* and ϕ_m is the *phase margin*. These margins serve as indicators as to how far the feedback system is from instability. As such, the gain and phase margins are useful in robustness analysis of control systems.

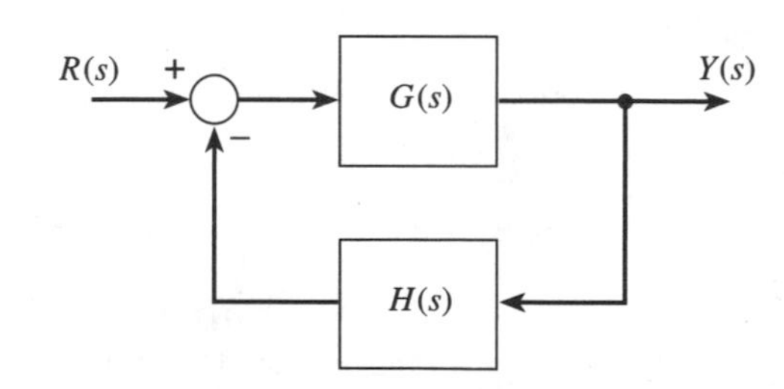

FIGURE 6.4 *Block diagram of a feedback system*

Although k_m and ϕ_m can be found graphically from the frequency-response plots, the Control System Toolbox provides the function `margin` which does this using an interpolation procedure. If $GH(s)$ is available as the LTI object `GH`, the gain and phase margins, the *gain-crossover frequency* (ω_g), and the *phase-crossover frequency* (ω_p) of the feedback system in Figure 6.4 can be computed using the command `[km,pm,wp,wg] = margin(GH)`. If the output variables are omitted, `margin` will automatically generate a Bode plot with the margins and crossover frequencies indicated. The following example illustrates a typical frequency-domain stability analysis.

EXAMPLE 6.4
Crossover Frequencies, Margins, and Stability

Use the function `margin` to find the gain and phase margins and the crossover frequencies of the feedback system of Figure 6.4 with $G(s)$ of Example 6.2 and $H(s) = 1$. Use the margins to determine the stability of the feedback system. Verify your answer by using the Nichols and Nyquist plots in Figures 6.2(b) and 6.2(c) and by computing the closed-loop system poles.

Solution

The MATLAB commands in Script 6.4 will compute the margins and crossover frequencies and generate the Bode plot shown in Figure 6.5. The `margin` function gives the gain margin $k_m = 29.86 = 29.50$ dB, the phase margin $\phi_m = 72.9°$, the phase-crossover frequency $\omega_p = 39.91$ rad/s, and the gain-crossover frequency $\omega_g = 0.904$ rad/s. Note that the phase-crossover frequency ω_p occurs when $\arg[GH(j\omega_p)] = -180°$, such that $k_m = 1/|GH(j\omega_p)|$ or $20 \log_{10}(1/|GH(j\omega_p)|)$ dB can be computed. The gain-crossover frequency ω_g occurs when $|GH(j\omega_g)| = 1$, from which $\phi_m = 180° + \arg[GH(j\omega_g)]$ can be computed.

MATLAB Script

```
% Script 6.4: gain and phase margin
numG = [1280 640]                % create GH(s) as TF object
denG = [1 24.2 1604.81 320.24 16]
G = tf(numG,denG)
H = 1                            % create H(s) as TF object (unity gain)
GH = G*H                         % open-loop system
margin(GH)                       % gain & phase margins and crossover freq
T = feedback(G,H)                % unity-feedback closed-loop system
poles_T = pole(T)                % closed-loop poles
```

Since $GH(s)$ is stable, the closed-loop system is stable because $k_m > 1$ and $\phi_m > 0$. For the Nichols plot in Figure 6.2(b), this stability condition is

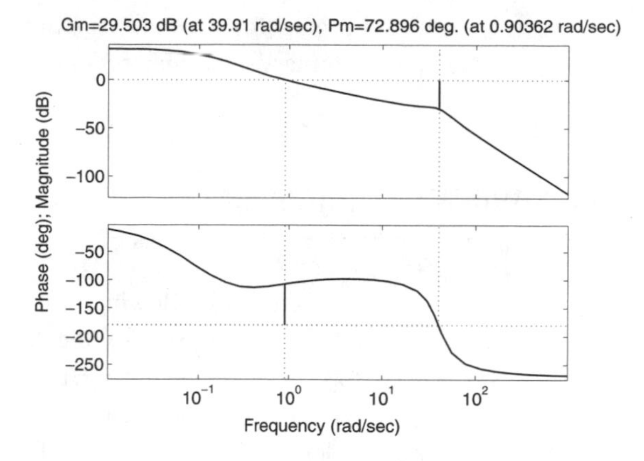

FIGURE 6.5 *Bode plot with margins and crossover frequencies for Example 6.4*

equivalent to the magnitude-phase curve being on the right of the point (0 dB, $-180°$). For the Nyquist plot in Figure 6.2(c), the frequency response curve does not encircle the critical point $(-1, 0)$. Since $GH(s)$ is stable, the Nyquist criterion ensures that the closed-loop system is stable. From the command `pole(T)`, the closed-loop system poles are at $s = -0.503 \pm j0.402$ and -11.60 ± 38.03, providing an independent verification of the stability.

Comment: We emphasize that frequency-domain stability criteria and control-design procedures are based on the frequency response of the *open-loop* transfer function $GH(s)$ and not on the closed-loop transfer function. This is the case in Example 6.4 where the margin calculation is based on $GH(s)$.

WHAT IF?

a. Repeat Example 6.4 by replacing $G(s)$ with $KG(s)$ where the gain K is set equal to k_m. You will find the new gain margin to be 1, the new phase margin to be zero, and the crossover frequencies to be equal; that is, $\omega_g = \omega_p$. This indicates that the closed-loop system is marginally stable as it has a pair of poles on the imaginary axis of the s-plane, with all the other poles in the left half-plane. Verify this observation by computing the closed-loop system poles, and show that the marginally-stable poles are at $s = \pm j\omega_g$.

b. Repeat Example 6.4 by replacing $G(s)$ with the compensated system $G_c(s)G(s)$, where $G_c(s)$ is a lead controller with the transfer function $(10s + 5)/(s + 5)$. Note the changes in the gain and phase margins. ■

CROSS-CHECK As a cross-check for the first part of the what-if question, perform a root-locus analysis of the feedback system. Show that the gain for marginal stability K^* is equal to the gain margin k_m, and the imaginary-axis intersections with the root locus occur at $\pm j\omega_g$. ■

REINFORCEMENT PROBLEMS

In each of the following problems, use the function `margin` to find the gain and phase margins for the feedback system in Figure 6.4, given the open-loop transfer function $GH(s)$. Draw the Bode, Nichols, and Nyquist plots for $GH(j\omega)$. For an open-loop stable system, use the gain and phase margins to determine the stability of the closed-loop system, and verify your result with the Nichols and Nyquist plots. For an open-loop unstable system, use the Nyquist stability criterion only. Verify the stability by computing the poles of the closed-loop system. As a cross check, perform a root-locus analysis and relate the gain for marginal stability K^* to the gain margin k_m.

P6.12 Open-loop stable system.

$$GH(s) = \frac{654(s+5)}{(s+1)(s+3)(s+12)(s+20)}$$

P6.13 Open-loop unstable system.

$$GH(s) = \frac{100(s+5)}{(s-2)(s+8)(s+20)}$$

Because $GH(s)$ is unstable, the Nyquist criterion has to be used for testing stability.

P6.14 High-gain system.

$$GH(s) = \frac{1000(s+8)}{s(s+2)(s^2+8s+32)}$$

EXPLORATION

E6.1 Interactive Exploration. The file `one_blk.m` can be used to make a variety of frequency-response plots for any single-block system in either the TF or SS form. The plotting menu includes making Bode, Nichols, and Nyquist plots. Use this capability to explore the effects on the frequency response (and thus on the closed-loop system stability) of:

a. an open-loop zero in the right half-plane
b. an open-loop pole in the right half-plane

c. a double pole at the origin of the s-plane
d. a zero at $s = 0$
e. a pair of lightly damped poles with a damping ratio of $\zeta = 0.1$
f. a notch filter

COMPREHENSIVE PROBLEMS

CP6.1 Electric power generation system. Run the file `epow.m` to obtain the transfer function $G_V(s)$ from $U(s)$ to $V_{\text{term}}(s)$. Determine its poles and zeros, and draw the Bode, Nichols, and Nyquist plots. Find the gain and phase margins of the system under unity feedback. Verify the gain margin k_m with the gain for marginal stability K^* found from the root-locus analysis in Comprehensive Problem CP5.1.

Repeat the analysis for the transfer function $G_\omega(s)$ from $U(s)$ to $\omega(s)$. You should notice that the presence of the electromechanical mode is more conspicuous in $G_\omega(s)$ than in $G_V(s)$.

CP6.2 Satellite. Use the file `sat.m` to create the transfer function from the motor torque to the pointing angle in TF form. Then use the result to obtain the frequency response as a Bode plot (use both the `bode` and `margin` functions), as a Nyquist plot, and as a Nichols plot. To get realistic values, include a gain of -0.001 in series with the transfer function (a negative controller gain is required with this system due to the way in which the variables are defined). You should find that the gain margin is infinite (the phase angle does not cross the $-180°$ line) and the phase margin is 6.5° when the gain value given above is used.

CP6.3 Stick balancer with rigid stick. Use the file `rigid.m` to obtain the transfer function from the motor voltage to the stick angle in TF form. Then draw a Bode plot of the open-loop frequency response. Because this model has a number of unusual features, including a double zero and a single pole at $s = 0$, a negative sign in the numerator, and a pole in the right half-plane, the gain and phase plots will differ from those of more ordinary systems. For example, the phase angle will approach $-270°$ for low frequencies and $-360°$ for high frequencies. Neither the gain margin nor the phase margin is defined because the gain remains below 0 dB and the phase angle remains below $-180°$ for all frequencies. Generate the Nyquist and Nichols plots and relate them to the behavior of the gain and phase in the Bode plot.

CP6.4 Hydro-turbine system. Run `hydro.m` to obtain the transfer function of the hydro-turbine system including the actuator model. Make a Bode plot of the model and observe that, although there are one zero and two poles, the phase of the frequency response goes to $-270°$ as $\omega \to \infty$. This is due to the presence of the non-minimum phase zero. Find the gain and phase margins of the feedback system in Figure A.8 in Appendix A with a unity-gain proportional controller. Also make the Nyquist and Nichols plots to see the effect of the non-minimum phase zero on the frequency-response plots.

SUMMARY

In this chapter we have illustrated the use of MATLAB to compute and plot frequency response. The frequency-response commands are applicable to systems in either the transfer-function or state-space form. Stability analysis in the frequency domain can be readily performed by using MATLAB to compute and plot the open-loop system frequency response. In particular, phase and gain margins give an indication of the stability robustness of a feedback system.

MATLAB FUNCTIONS USED

Function	*Purpose and Use*	*Toolbox*
abs	Given a scalar, vector, or matrix, **abs** returns its absolute value.	MATLAB
bode	Given a model in TF or SS form, **bode** returns the magnitude and phase of the frequency response. When the output variables are omitted, it generates the Bode plot directly.	Control System
damp	Given an LTI object, **damp** returns the natural frequencies and damping factors of the system.	Control System
dcgain	Given a model in TF or SS form, **dcgain** returns the steady-state gain of the system.	Control System
feedback	Given the models of two systems in TF form, **feedback** returns the model of the closed-loop system, where negative feedback is assumed. An optional argument can be used to handle the positive feedback case.	Control System
logspace	The function **logspace** generates vectors whose elements are logarithmically spaced.	MATLAB
lsim	Given an LTI object of a continuous system, a vector of input values, a vector of time points,	Control System

	and possibly a set of initial conditions, **lsim** returns the time response.	
margin	Given a model in TF or SS form, **margin** returns the gain and phase margins and the crossover frequencies. When the output variables are omitted, it generates a Bode plot, with the margins and crossover frequencies indicated on the plot.	Control System
ngrid	The function **ngrid** generates grid lines for a Nichols plot.	Control System
nichols	Given a model in TF or SS form, **nichols** returns the magnitude and phase of the frequency response. When the output variables are omitted, it generates the Nichols plot directly.	Control System
nyquist	Given a model in TF or SS form, **nyquist** returns the real and imaginary parts of the frequency response. When the output variables are omitted, it generates the Nyquist plot directly.	Control System
pole	Given an LTI object, **pole** computes the poles of the system's transfer function.	Control System
reshape	Given a multidimensional array, **reshape** can be used to change its dimensions.	MATLAB
semilogx	The function **semilogx** generates semi-logarithmic plots, using a base 10 logarithmic scale for the x-axis and a linear scale for the y-axis.	MATLAB
tf	Given numerator and denominator polynomials, **tf** creates the system model as a TF object. The command also converts zero-pole-gain or state-space models to TF form.	Control System
zpkdata	Given an LTI object, **zpkdata** extracts the zeros, poles, and gain and other information about the system.	Control System

ANSWERS

P6.1 $y_{ss}(t) = 3.571 \cos(2t - 138.11°)$

P6.2 $y_{ss}(t) = 0.995 \sin(0.1t - 5.14°) + 0.711 \cos(t + 5.71°) - 0.141 \sin(10t - 69.3°)$

P6.3 $y_{ss}(t) = 0.101 \sin(0.1t + 5.14°) + 0.141 \cos(t + 84.3°) - 0.711 \sin(10t + 9.29°)$

P6.4 $y_{ss}(t) = 8$

P6.5 $y_{ss}(t) = 1.99 \cos(2t - 95.7°)$

P6.12 $k_m = 10$, $\phi_m = 67.6°$, $\omega_p = 14.6$ rad/s, $\omega_g = 3.3$ rad/s; closed-loop system stable

P6.13 $k_m = 0.64$, $\phi_m = 55.1°$, $\omega_p = 0.04$ rad/s, $\omega_g = 2.66$ rad/s; closed-loop system stable

P6.14 $k_m = 0.053$, $\phi_m = -69.5°$, $\omega_p = 3.4$ rad/s, $\omega_g = 10.6$ rad/s; closed-loop system unstable

System Performance

PREVIEW

The main objectives of feedback control system design are to guarantee system stability and to satisfy certain desired performance criteria. Performance can be measured in either the time domain or the frequency domain. Typical time-domain measures include the damping ratio of the dominant poles and step-response rise time, overshoot, settling time, and steady-state command tracking and disturbance rejection. In the frequency domain, typical measures include the gain and phase margins, DC gain, and closed-loop system bandwidth. In addition, sensitivity functions can be generated to evaluate the effect of gain variations on the feedback system. Approximate relationships between some of the time-domain and frequency-domain measures can be found in tables and charts in many control systems textbooks. The purpose of this chapter is to use MATLAB to compute system performance measures and to understand how the controller parameters may affect the performance. These performance measures will be used in the remaining chapters where we deal with the controller design.

TIME-DOMAIN PERFORMANCE

Consider the feedback system in Figure 7.1(a) where $G_p(s)$ is the process or plant model, $H(s)$ the sensor dynamics in the feedback path, and $G_c(s)$

the controller. The system has two inputs: the reference input $r(t)$ whose Laplace transform is $R(s)$, and the disturbance input $d(t)$ whose Laplace transform is $D(s)$. The other signals are: $y(t)$, with Laplace transform $Y(s)$, the process output; $\Delta(t)$, with Laplace transform $\Delta(t)$, the difference signal obtained by subtracting the sensor output from the reference signal; $m(t)$, the controller output, whose Laplace transform is $M(s)$; and $u(t)$, the input to the process that combines the controller output and the disturbance input, whose Laplace transform is $U(s)$. If $H(s) = 1$, then $\Delta(t)$ becomes the regulation error $e(t) = r(t) - y(t)$.

The objective of the controller is to make the process output, $y(t)$, be a good approximation to the reference or command input $r(t)$, while having $y(t)$ be relatively unaffected by the disturbance $d(t)$. To analyze the command tracking performance of the feedback system in Figure 7.1(a), we need the closed-loop transfer function from $R(s)$ to $Y(s)$ given by

$$T_r(s) = \frac{G_c(s)G_p(s)}{1 + G_c(s)G_p(s)H(s)} \tag{7.1}$$

From the simplified diagram in Figure 7.1(b) with the disturbance as the only input, the disturbance rejection performance can be obtained from the

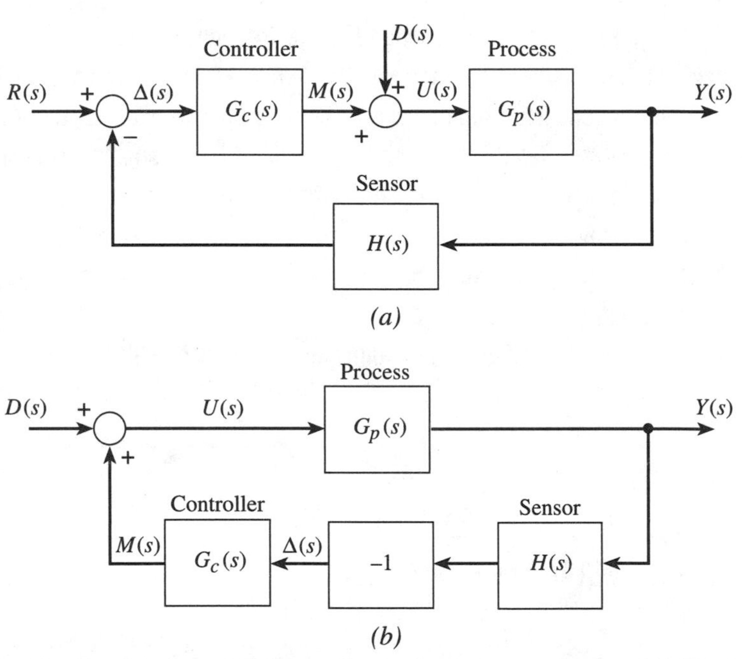

FIGURE 7.1 *Feedback system definitions for performance evaluation (a) Block diagram with two inputs (b) Simplified diagram for analysis of disturbance input*

closed-loop transfer function from $D(s)$ to $Y(s)$ given by

$$T_d(s) = \frac{G_p(s)}{1 + G_c(s)G_p(s)H(s)} \tag{7.2}$$

In the time domain, the command-tracking performance of the controller is commonly measured in terms of how well the closed-loop system's response follows a step reference input. The performance measures include: t_r, the rise time for $y(t)$ to go from 10% to 90% of the reference value $r(t) = y_{\text{ref}}$; M_o, the maximum overshoot in percent defined as

$$M_o = \frac{y_{\text{peak}} - y_{\text{ref}}}{y_{\text{ref}}} \times 100$$

where y_{peak} is the peak value of $y(t)$; t_{s2}, the settling time for $y(t)$ to remain within 2% of y_{ref}; and e_{ss}, the steady-state regulation error.

The RPI function `tstats` has been created to compute these performance measures. The arguments of `tstats` are: (i) a column vector of time values, (ii) a corresponding column vector of the step response to be evaluated, and (iii) a scalar specifying the reference value of $y(t)$. The use of `tstats` is illustrated in the following example.

SYSTEM PERFORMANCE

EXAMPLE 7.1 ***Step-Response Performance***

Consider the feedback system in Figure 7.1 where

$$G_p(s) = \frac{1}{(s + 0.01)(s + 1)(s + 20)}, \qquad G_c(s) = 50 \quad \text{and} \quad H(s) = 1$$

Simulate the reference response $y_r(t)$ due to a unit-step input $r(t)$ with $d(t) = 0$ and use `tstats` to find M_o, t_p, t_r, t_{s2}, and e_{ss}. Also find the damping ratios and the undamped natural frequencies of the closed-loop system poles. Then simulate the response $y_d(t)$ due to a unit-step disturbance $d(t)$ with $r(t) = 0$ and find its steady-state value.

Solution

The MATLAB commands to perform the required computations are given in Script 7.1. First, the closed-loop system transfer function

$$T_r(s) = \frac{50}{s^3 + 21.01s^2 + 20.21s + 50.2}$$

is created as the LTI object `Tr`. Then the column vector of time `t` is set up and the `step` command is used to generate the unit-step response $y(t)$, which is shown in Figure 7.2(a). Applying the function `tstats` to the unit-step response with $y_{\text{ref}} = 1$, we find the following performance measures: the rise time t_r is 0.83 s; the maximum overshoot M_o is 39.5% and occurs at $t_p = 2.1$ s; the 2% settling time t_{s2} is 9.0 s; and the steady-state error e_{ss} is 0.4%.

MATLAB Script

```
% Script 7.1: Step-response performance
denGp = conv(conv([1 0.01],[1 1]),[1 20])      % plant model
Gp = tf(1,denGp)
Gcp = 50*Gp                                     % controller & plant
Tr = feedback(Gcp,1.0)                          % unity feedback
t = [0:0.05:20]';                               % time vector
yr = step(Tr,t);                                % closed-loop step response
plot(t,yr);grid                                 % figure 7.2(a)
[Mo,tp,tr,ts2,ess] = tstats(t,yr,1)             % performance measures
damp(Tr)                                        % CL pole damping ratio
[numTr,denTr] = tfdata(Tr,'v')                  % extract polynomials
[resS,polS,otherS] = residue(numTr,[denTr 0])   % partial-fraction expansion
Td = feedback(Gp,50)                            % compute Td(s)
step(Td,t)                                      % disturbance step response
```

CHAPTER 7

The closed-loop system poles, and their damping ratios and natural frequencies are obtained with the damp command, and are listed in Table 7.1. Since the damping ratio of the poles at $s = -0.4401 \pm 1.517$ is small, due to the small negative real part, we conclude that they are the dominant poles of the closed-loop system. The pole at $s = -20.13$ is due to a small time constant and has a small residue component. Thus it is not dominant.

TABLE 7.1 *Damping ratios and natural frequencies of closed-loop system poles*

Pole	*Damping ratio*	*Natural freq.*
-20.13	1.00	20.13
-0.4401 ± 1.517	0.28	1.579

The disturbance transfer function $T_d(s)$, as expressed in (7.2), is used to generate the unit-step disturbance response, as shown in Figure 7.2(b), where the steady-state value is 0.02.

Comment: If $G_c(s)$ is a constant gain K_P, then it follows from (7.1) and (7.2) that $T_d(s)$ can be obtained by scaling $T_r(s)$ with K_P; that is, $T_d(s) = T_r(s)/K_P$. This is evident from Figure 7.2, which shows that $y_d(t) = y_r(t)/50$.

WHAT IF?

a. Redo Example 7.1 using $G_c(s) = 100$ as the controller. You will find that the rise time and the steady-state error are smaller. However, the overshoot is larger, due to a smaller damping ratio, and the settling time is longer.

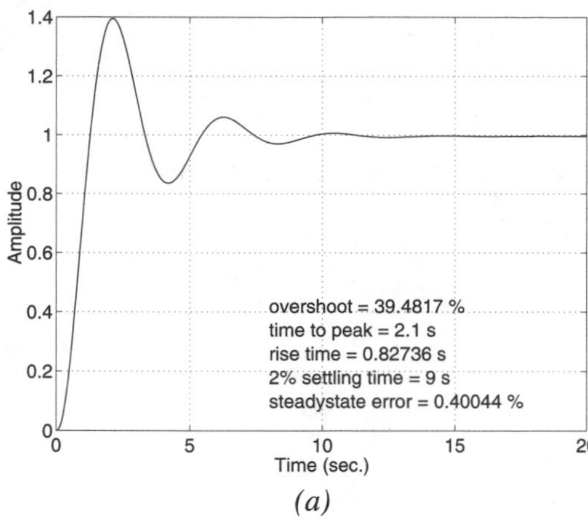

(a)

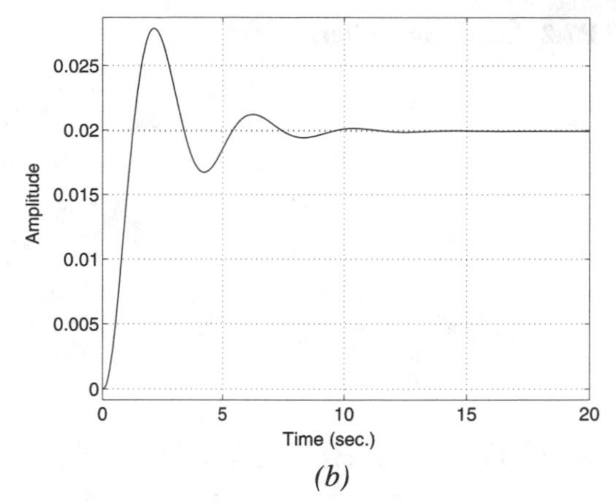

(b)

FIGURE 7.2 *Step responses for Example 7.1 (a) Command response (b) Disturbance response*

b. Redo Example 7.1 using a lag controller with the transfer function

$$G_c(s) = \frac{50(s + 0.2)}{s + 0.02}$$

You will find that the steady-state error is smaller but the settling time is longer. All the other performance measures remain about the same. In addition, $y_d(t)$ can no longer be obtained from $y_r(t)$ by dividing it with a constant. ■

REINFORCEMENT PROBLEMS

In each of the following problems, the plant $G_p(s)$ and the controller $G_c(s)$ for the feedback system in Figure 7.1 are specified. Assume $H(s) = 1$ in all the problems. Simulate the reference response $y_r(t)$ due to a unit-step input $r(t)$ and find M_o, t_p, t_r, t_{s2}, and e_{ss}. Also find the damping ratios and the natural frequencies of any complex closed-loop poles. Then simulate the response $y_d(t)$ due to a unit-step disturbance $d(t)$, and find its steady-state value.

P7.1 Plant with an integrator.

$$G_p(s) = \frac{1}{s(s + 1)(s + 20)} \quad \text{and} \quad G_c(s) = 50$$

The system is obtained from the system in Example 7.1 by replacing the pole at $s = -0.01$ with the pole at $s = 0$. Verify that e_{ss} due to the unit-step input $r(t)$ will be zero.

P7.2 Lead controller.

$$G_p(s) = \frac{1}{s(s+1)(s+20)} \quad \text{and} \quad G_c(s) = \frac{500(s+0.5)}{s+5}$$

Compared to the proportional controller in Problem 7.1, the lead controller will achieve faster rise and settling times.

P7.3 Lightly damped mode.

$$G_p(s) = \frac{s+1}{(s+0.1)(s^2+3s+25)} \quad \text{and} \quad G_c(s) = 100$$

STEADY-STATE REGULATION

Consider the feedback control system given in Figure 7.1 with $H(s) = 1$. In analyzing the steady-state tracking of the reference input $r(t)$ by the plant output $y_r(t)$, it is useful to introduce the concept of system type. The open-loop system $G_c(s)G_p(s)$ is classified as a type-N system if $G_c(s)G_p(s)$ has N poles at the origin. For a type-0 system, we define the position error constant as

$$K_p = \lim_{s \to 0} G_c(s)G_p(s) \tag{7.3}$$

If the closed-loop system is asymptotically stable, then the steady-state error for a step input $r(t) = A$ is $e_{ss} = A/(1 + K_p)$. For a type-1 system, we define the velocity error constant as

$$K_v = \lim_{s \to 0} sG_c(s)G_p(s) \tag{7.4}$$

If the closed-loop system is asymptotically stable, then the steady-state error for a ramp input $r(t) = Bt$ is $e_{ss} = B/K_v$.

To compute K_p, the Control System Toolbox provides the function `dcgain`, whose usage has already been illustrated in several examples. However, it does not have a function to directly compute K_v. As a result, we have created the RPI function `vgain` for this purpose. Given the LTI object `G`, the command `Kv = vgain(G)` will return the velocity error constant K_v. The next example illustrates the calculation of K_v.

EXAMPLE 7.2 ***Velocity Error Constant***

Find K_v for the type-1 feedback system in Problem 7.1, and simulate the unit-ramp response of the closed-loop system.

Solution

In Script 7.2, after the series connection $G(s) = G_c(s)G_p(s)$ has been computed, the function `vgain` is used to find the velocity error constant $K_v = 2.5$.

Thus the steady-state error due to a unit-ramp input is $1/2.5 = 0.40$ and is verified by the unit-ramp response of the closed-loop system shown in Figure 7.3.

MATLAB Script

```
% Script 7.2:  velocity error constant calculation
denGp = conv(conv([1 0],[1 1],[1 20])  % multiply 3 polynomials
Gp = tf(1,denGp)                       % generate Gp(s)
G = 50*Gp                              % series connection
Kv = vgain(G)                          % velocity error constant
T = feedback(G,1)                      % unity feedback system
t = [0:0.05:10]'; u = t;               % time and unit ramp
y = lsim(T,u,t);                       % ramp response
plot(t,u,'-',t,y,'--')                 % Figure 7.3
```

FIGURE 7.3 *Plot of unit-ramp input and response for Example 7.2*

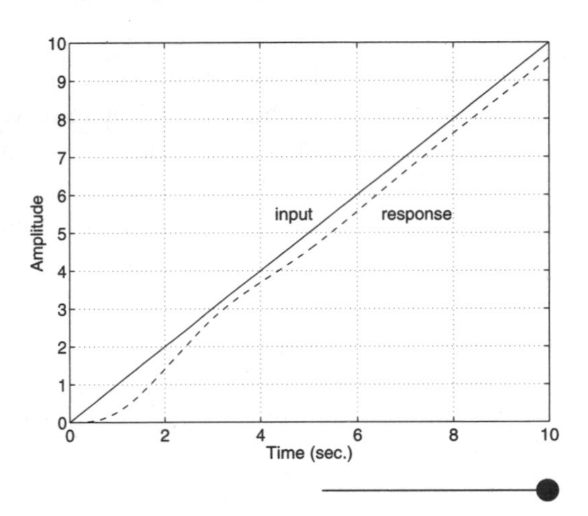

REINFORCEMENT PROBLEMS

In each of the following problems, consider the command tracking by the plant output for the feedback system in Figure 7.1. The plant and controller transfer functions $G_p(s)$ and $G_c(s)$ are given and $H(s) = 1$ is assumed. Verify the system type number and find the appropriate steady-state error constant. Simulate the unit-step and unit-ramp responses of the closed-loop system.

P7.4 Type-0 system.

$$G_p(s) = \frac{6(s + 10)}{(s + 3)(s^2 + 4s + 4)} \quad \text{and} \quad G_c(s) = 3$$

P7.5 Type-1 system.

$$G_p(s) = \frac{6(s+1)}{s(s^2+4s+4)} \quad \text{and} \quad G_c(s) = 5$$

FREQUENCY-DOMAIN PERFORMANCE

Open- and closed-loop system frequency responses provide additional means of measuring the performance of feedback systems. The gain and phase margins computed from the frequency response of the open-loop transfer function $G_c(s)G_p(s)H(s)$ provide an indication of the robustness of feedback system. In addition, we can also measure the system's performance by using the frequency response of the closed-loop transfer function $T_r(s)$ (7.1) from $R(s)$ to $Y(s)$. The *bandwidth* of the closed-loop system is defined to be the frequency ω_B where $|T_r(j\omega_B)| = (1/\sqrt{2})\,|T_r(0)| = 0.707|T_r(0)|$; that is, $|T_r(j\omega_B)|$ is 3 dB below the DC gain magnitude $|T_r(0)|$. The bandwidth serves as an indication of the speed of the response of a system to inputs. The higher the bandwidth of a system, the faster is its response.

For a system with a single pair of dominant complex poles having its frequency response peak greater than $|T_r(0)|$, we define the resonance peak as

$$M_p = \max_{\omega} |T_r(j\omega)| \tag{7.5}$$

and ω_p as the resonance frequency where $|T_r(j\omega_p)| = M_p$. For a second-order system with no zeros, M_p can be related to the phase margin and damping ratio. In general, a smaller phase margin and damping ratio will lead to a higher value of M_p.

One purpose of a feedback system is to reduce the effects of parameter variations on the system performance. However, this reduction is not uniform over all frequencies. To gain further insight, we separate a gain K from the controller transfer function by writing $G_c(s) = KF_c(s)$. Then the change of $T_r(s)$ with respect to a change of K is defined as the sensitivity function

$$S(s) = \frac{\partial T_r(s)/T_r(s)}{\partial K/K} = \frac{1}{1 + KF_c(s)G_p(s)H(s)} = \frac{1}{1 + G_c(s)G_p(s)H(s)} \tag{7.6}$$

We can show that $S(s)$ is also the transfer function from $R(s)$ to $\Delta(s)$. The frequency response $S(j\omega)$ can be used to evaluate the robustness of the feedback system with respect to a variation in the controller gain. A large value of $|S(j\omega)|$ indicates that the characteristics of the closed-loop system (magnitude and phase) at the frequency ω will be sensitive to the variations of K.

The following example illustrates the computation of these frequency-domain performance measures. To compute the margins and the corresponding crossover frequencies, we use the Control System Toolbox function `margin`. To compute the closed-loop bandwidth, we have created the RPI function `bwcalc` which has three arguments: (i) the frequency points in the column vector `w`, (ii) the corresponding magnitude $T_r(j\omega)$ in decibels in the column vector `magTr_dB`, and (iii) the low frequency gain $|T_r(0)|$ in decibels in the scalar `lfgTr_dB`. The command `bw = bwcalc(magTr_dB,w,lfgTr_dB)` returns the closed-loop bandwidth as the scalar `bw`. The variables M_p and ω_p can be obtained by applying the MATLAB function `max` to the vector `magTr_dB`. In the following example we will illustrate the evaluation of these frequency-response measures.

EXAMPLE 7.3 ***Frequency-Response Performance***

For the feedback system given in Example 7.1, make a Bode plot for the open-loop transfer function $G(s) = G_c(s)G_p(s)$, and find the gain and phase margins and the related crossover frequencies of the feedback system. Then plot $|T_r(j\omega)|$ and find ω_B, M_p, and ω_p for the closed-loop system. Finally, plot $|S(j\omega)|$ to investigate the closed-loop system sensitivity with respect to the controller gain.

Solution

The commands to compute the performance measures are contained in Script 7.3. After the transfer functions $G_p(s)$ and $G_c(s)$ are entered ($H(s) = 1$), the $*$ function is used to form the open-loop transfer function $G(s)$. The function `margin` is applied to $G(s)$ to find the gain margin $k_m = 18.6$ dB and the phase margin $\phi_m = 31.3°$. The gain- and phase-crossover frequencies are 1.43 and 4.50 rad/s, respectively. A Bode plot showing these margins and the crossover frequencies is given in Figure 7.4.

MATLAB Script

```
% Script 7.3:  Frequency response performance
denGp = conv(conv([1 0.01],[1 1]),[1 20]) % generate Gp(s)
Gp = tf(1,denGp)
G = 50*Gp                                % series connection
margin(G)                                % figure 7.4
Tr = feedback(G,1)                       % build CL transfer function T(s)
bode(Tr)                                 % plot CL frequency response
[mag_CL,ph_CL,w] = bode(Tr);             % CL gain as ratio & phase
mag_CLdB = 20*log10(mag_CL);             % gain in dB
lfg_CL = dcgain(Tr)                      % dc gain as ratio
lfg_CL_dB = 20*log10(lfg_CL);            % dc gain in decibels
BW = bwcalc(mag_CLdB,w,lfg_CL)           % closed-loop bandwidth in rad/s
```

Continues

SYSTEM PERFORMANCE

```
[M_pw,i] = max(mag_CLdB)                    % maximum CL frequency respresponse
wp = w(i)                                   % .....at this frequency (rad/s)
S = feedback(1,G)                           % sensitivity function S(s)
[magS,phS] = bode(S,w);                     % sensitivity frequency response
magS_dB = 20*log10(magS);                   % sensitivity magnitude in dB
[maxS,j] = max(magS_dB)                     % maximum sensitivity magnitude
w(j)                                        % .....at this frequency (rad/s)
semilogx(w,mag_CLdB(:),'-',w,magS_dB(:),'--') % figure 7.5
```


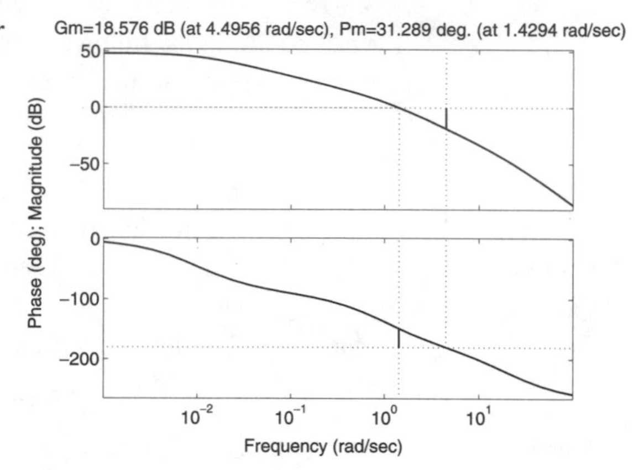

FIGURE 7.4 *Bode plot for Example 7.3 indicating margins and crossover frequencies*

The closed-loop transfer function $T_r(s)$ is computed from the function `feedback` and is given in Example 7.1. The magnitude $|T_r(j\omega)|$ obtained from the function `bode` is converted to decibels in the column array `mag_CLdB` and is plotted versus ω as the solid curve in Figure 7.5. The low frequency gain `lfg_CL` is computed from the function `dcgain` and is converted to decibels in `lfg_CL_dB`. Together with the frequency array `w`, these properties are used by `bwcalc` to compute the closed-loop bandwidth as $\omega_B = 2.32$ rad/s. From `mag_CLdB` and `w`, we also obtain $M_p = 5.25$ dB and $\omega_p = 1.53$ rad/s. Note that the `max` command returns both the peak value in `Mp` and its index in `i`. Hence `w(i)` is the frequency at which the peak value occurred.

Treating $G_c(s)G_p(s)$ as the transfer function in the feedback path, the function `feedback` is used to generate the sensitivity function (7.6)

$$S(s) = \frac{s^3 + 21.01s^2 + 20.21s + 0.2}{s^3 + 21.01s^2 + 20.21s + 50.2}$$

Its magnitude $|S(j\omega)|$ is plotted versus ω as the dashed curve in Figure 7.5. At low frequencies, the sensitivity magnitude is small, which indicates that the low-frequency closed-loop system performance is not greatly affected by a variation in the controller gain. However, $|S(j\omega)|$ peaks at 1.8 rad/s with a value of 6.45 dB, indicating that the closed-loop system performance in the neighborhood of 1.8 rad/s is sensitive to the controller gain variation. This is because the frequency of the dominant closed-loop mode is 1.52 rad/s, and this mode is strongly affected by the controller gain. At high frequencies, $|T_r(j\omega)|$ approaches zero and $|S(j\omega)|$ approaches unity.

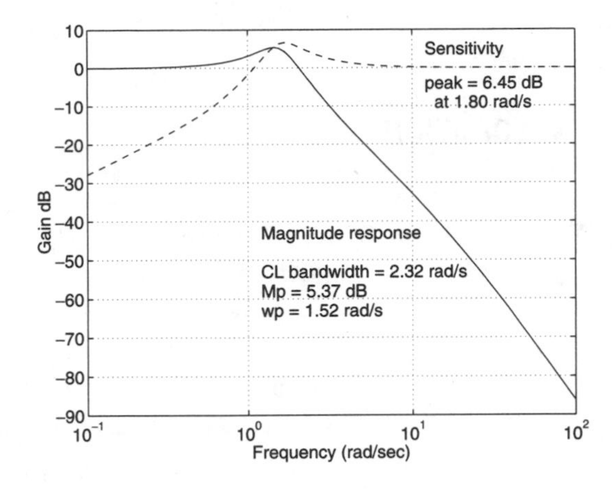

FIGURE 7.5 *Magnitude plots of $T(j\omega)$ and $S(j\omega)$ for Example 7.3*

WHAT IF? Redo Example 7.3 using the proportional controller $G_c(s) = 100$. You will find that the margins will be smaller and the closed-loop bandwidth and the resonance peak will be higher. In addition, the peak of the sensitivity function magnitude will be higher and occur at a higher frequency. ■

REINFORCEMENT PROBLEMS

In each of the following problems, the plant $G_p(s)$, the controller $G_c(s)$, and the feedback-path transfer function $H(s)$ for Figure 7.1 are given. Make a Bode plot for the open-loop transfer function $G_c(s)G_p(s)H(s)$, and find the gain and phase margins and the related crossover frequencies. Plot the magnitude of the closed-loop frequency response $|T_r(j\omega)|$, and find the closed-loop bandwidth ω_B, the resonance peak M_p, and the resonance frequency ω_p. Also plot the magnitude of the closed-loop sensitivity function $|S(j\omega)|$, and comment on the closed-loop system sensitivity with respect to the controller gain. Furthermore, find the damping ratio ζ of the dominant poles of the closed-loop system. Simulate the unit-step response of the closed-loop system, and find the rise time t_r and the maximum overshoot M_o in percent.

P7.6 Proportional Controller.

$$G_p(s) = \frac{s+5}{s^2+13s+12}, \qquad G_c(s) = 15 \quad \text{and} \quad H(s) = \frac{60}{(s+3)(s+20)}$$

P7.7 PI Controller.

$$G_p(s) = \frac{1}{(s^2+8s+32)}, \qquad G_c(s) = \frac{50(s+8)}{s} \quad \text{and} \quad H(s) = 1$$

CHAPTER 7

EXPLORATION

E7.1 Effects on System Performance. The file one_blk.m has an option in its menu that allows the user to implement a unity-feedback system. After the model of a system has been entered, the user can compute the closed-loop system transfer function and the sensitivity function and make various Bode plots. Use this option to explore the effects on closed-loop system performance in the time and frequency domains of plants with

a. an open-loop pole at $s = 0$
b. an open-loop zero at $s = 0$
c. an open-loop zero in the right half-plane
d. an open-loop pole in the right half-plane
e. a pair of lightly damped open-loop poles with $\zeta < 0.1$

COMPREHENSIVE PROBLEMS

CP7.1 Electric power generation system. Run the file epow.m to obtain the transfer function of the power system from u to V_{term}. Investigate the voltage-control system in Figure A.2 in Appendix A using a lag controller (3.2) with $\alpha_{\text{lag}} = 10$ for several values of the controller gain K_{lag} and of the zero location z_{lag}. For each choice of K_{lag} and z_{lag}, simulate the closed-loop system response $V_{\text{term}}(t)$ to a unit-step reference voltage $V_{\text{ref}}(t)$. Find M_o, t_p, t_r, t_{s2}, and e_{ss}. Also find the damping ratios and the natural frequencies of the closed-loop system poles. In addition, make Bode plots for the closed-loop transfer function $T_r(s)$ (7.1) and the sensitivity function $S(s)$ (7.6), and comment on the effect of the variation of K_{lag} on the feedback system.

CP7.2 Satellite with TF model (zero initial wheel speed).

a. Run the file sat.m to obtain the transfer function of the satellite from the motor torque τ_m to the pointing angle θ. Then connect a lead controller (3.3) with $\alpha_{\text{lead}} = 10$ in series with the satellite model and obtain the transfer function of the closed-loop system with unity feedback. Use the resulting transfer function $T_r(s)$ to compute and plot the unit-step response for values of K_{lead} in the interval $[-0.0005, -0.002]$ and of z_{lead} in the interval $[-0.05, -0.01]$. In each case, use the RPI function `tstats`, with its third argument set to 1.0, to compute M_o, t_p, t_r, t_{s2}, and e_{ss}. Comment on the relationships of the controller gain and zero location on the various performance measures.

b. Use the open-loop model of the lead compensator in series with the satellite to establish whether the system is type-0 or type-1. Then compute the appropriate error constant (K_p if type-0 and K_v if type-1). Note that the numerator of the open-loop transfer function has one zero at $s = 0$ and the denominator has two zeros at $s = 0$. To compute the appropriate error constant, you must use the transfer function with the pole-zero cancellation taken into account. This can be accomplished by dividing both the numerator and denominator polynomials by s. You can do this in MATLAB with the `deconv` command. For instance, the command `num = deconv([3 2 0], [1 0])` yields `num = [3 2]`. Then form the unity-feedback closed-loop transfer function and use it to compute and plot the response to a unit-ramp reference input.

c. Using the transfer function with the zero and pole at the origin canceled, plot the open-loop frequency response and determine the gain and phase margins and the crossover frequencies. Then compute and plot the closed-loop frequency response $|T_r(j\omega)|$ and determine the low-frequency gain and the closed-loop system bandwidth. Finally, compute and plot the sensitivity function $|S(j\omega)|$ and comment on its behavior.

CP7.3 Hydro-turbine system. Run the file `hydro.m` to obtain the hydro-turbine system model. Apply a proportional controller with the gain $K_P = 0.5$ to the power control system in Figure A.8 in Appendix A, and find the closed-loop transfer function $T_r(s)$ and the sensitivity function $S(s)$. From the closed-loop system frequency response, find the bandwidth and the magnitude and frequency of the resonance peak. Comment on the sensitivity of the closed-loop system to the variation in the controller gain K_P. Then plot the response of the closed-loop system to a unit-step reference input. Comment on why it would not be appropriate to apply the function `tstats` to the step response to find the rise time and the settling time.

COMPREHENSIVE PROBLEMS

SUMMARY

We have summarized a number of performance measures for feedback control systems in both the time and frequency domain and shown how MATLAB can be used to evaluate and display them. To assist in the evaluation of the performance measures, we have created RPI functions to find the velocity error constant, the step-response measures, and the bandwidth. The time-domain performance measures will be used in Chapter 8 to design PID controllers and in Chapter 10 to design state-space controllers, and those for the frequency-domain will be used in Chapter 9 to design lead-lag compensators.

MATLAB FUNCTIONS USED

Function	*Purpose and Use*	*Toolbox*
*	Given two LTI objects, the * operator forms their series connection.	Control System
bode	Given a model in TF or SS form, **bode** returns the magnitude and phase of the frequency response. When the output variables are omitted, it generates the Body plot directly.	Control System
bwcalc	Given the magnitude of the frequency response of a system and its low frequency gain, **bwcalc** computes the bandwidth.	RPI function
conv	Given two row vectors containing the coefficients of two polynomials, **conv** returns a row vector containing the coefficients of the product of the two polynomials.	MATLAB
damp	Given a model as an LTI object, **damp** calculates the natural frequencies and the damping ratios of the system poles. When invoked without output variables, a table of poles, damping ratios, and natural frequencies is displayed.	Control System
dcgain	Given a model in TF or SS form, **dcgain** computes the position error constant.	Control System
deconv	Given two polynomials as row vectors, **deconv** returns the quotient and remainder of the first polynomial divided by the second.	MATLAB
feedback	Given the models of two systems in TF or SS form, **feedback** returns the model of the closed-loop system, where negative feedback is assumed.	Control System

	An optional third argument can be used to handle the positive feedback case.	
lsim	Given a TF or SS model of a continuous system, a vector of input values, a vector of time points, and possibly a set of initial conditions, **lsim** returns the time response.	Control System
margin	Given a model in TF or SS form, **margin** returns the gain and phase margins and the crossover frequencies. When the output variables are omitted, it generates a Bode plot, with the margins and crossover frequencies indicated on the plot.	Control System
residue	Given a rational function $T(s) = N(s)/D(s)$, **residue** returns the roots of $D(s) = 0$, the partial-fraction coefficients, and any polynomial term that remains.	MATLAB
step	Given a TF or SS model of a continuous system, **step** returns the response to a unit-step function input.	Control System
tf	Given numerator and denominator polynomials, **tf** creates the system model as a TF object. The command also converts zero-pole-gain or state-space models to TF form.	Control System
tfdata	Given a TF object, **tfdata** extracts the numerator and denominator polynomials and other information about the system.	Control System
tstats	Given a step response, **tstats** finds the percent overshoot, peak time, rise time, settling time, and steady-state error.	RPI function
vgain	Given a model as an LTI object, **vgain** computes the velocity error constant.	RPI function

ANSWERS

P7.1 $t_r = 0.82$ s, $M_o = 40.4\%$, $t_p = 2.1$ s, $t_{s2} = 8.85$ s, $e_{ss} = 0$, $\zeta = 0.276$, $\omega_n = 1.58$ rad/s, y_{ss} due to unit-step disturbance $= 0.02$

P7.2 $t_r = 0.34$ s, $M_o = 15.0\%$, $t_p = 0.75$ s, $t_{s2} = 3.4$ s, $e_{ss} = 0$, $\zeta = 0.405$, $\omega_n = 5.06$ rad/s, y_{ss} due to unit-step disturbance = 0.02

P7.3 $t_r = 0.19$ s, $M_o = 20.6\%$, $t_p = 0.4$ s, t_{s2} not applicable, $e_{ss} = 4.0\%$, $\zeta = 0.0507$, $\omega_n = 4.90$ rad/s, y_{ss} due to unit-step disturbance = 0.016

P7.4 $K_p = 15$, $K_v = 0$

P7.5 $K_p = \infty$, $K_v = 7.5$

P7.6 $k_m = 17.3$ dB, $\phi_m = 57.4°$, $\omega_B = 18.2$ rad/s, $M_p = 6.38$ dB, $\omega_p = 5.74$ rad/s, $\zeta = 52.6\%$, $t_r = 0.075$ s, $M_o = 54.8\%$

P7.7 $k_m = \infty$, $\phi_m = 18.7°$, $\omega_B = 11.5$ rad/s, $M_p = 9.90$ dB, $\omega_p = 8.17$ rad/s, $\zeta = 13.3\%$, $t_r = 0.16$ s, $M_o = 50.3\%$

Proportional-Integral-Derivative Control

PREVIEW

Now that we have become proficient in building system models and analyzing them with the aid of MATLAB and have established a number of measures for evaluating the performance of a feedback control system, we are ready to consider methods for designing control systems. A basic controller prototype is one that uses the error, its time derivative, and its integral with respect to time to construct the signal that is used to drive the actuators, which in turn affect the behavior of the process being controlled.

We will start by showing that using a control signal that is proportional to the error cannot be expected to result in good damping and fast response and may have unacceptable steady-state error. Introducing an integral term in the controller can eliminate steady-state errors but may adversely affect the damping. A term proportional to the derivative of the error can improve the speed of response and the damping, but it will not reduce steady-state errors. A controller that combines all three terms, known as a proportional-integral-derivative (PID) controller can provide significant improvements in response time, damping, and steady-state error reduction.

PROPORTIONAL CONTROL

Consider the feedback system shown in Figure 8.1(a) that was used in Chapter 7 to establish closed-loop system performance measures. In this chapter, we consider the use of proportional, integral, and derivative functions in the controller $G_c(s)$ to meet the system performance specifications. In addition to Figure 8.1(a), we will also use the diagram in Figure 8.1(b) to evaluate the response of a particular design to a disturbance input.

In this section we start with the design of a proportional controller with $G_c(s) = K_P$. In general, as the proportional gain K_P is increased, the rise time (t_r) and the steady-state regulation error for a unit-step reference input will become smaller. However, these performance improvements will be offset by a larger overshoot. Thus a typical design often involves some performance tradeoffs. For second-order systems with no zeros, formulas and charts found in many standard control textbooks can be used to obtain an appropriate K_P. However, for higher-order systems, formulas to obtain exact K_P are not available, requiring a designer to iteratively select an appropriate K_P. MATLAB is an ideal computer tool for performing this type of iterative design.

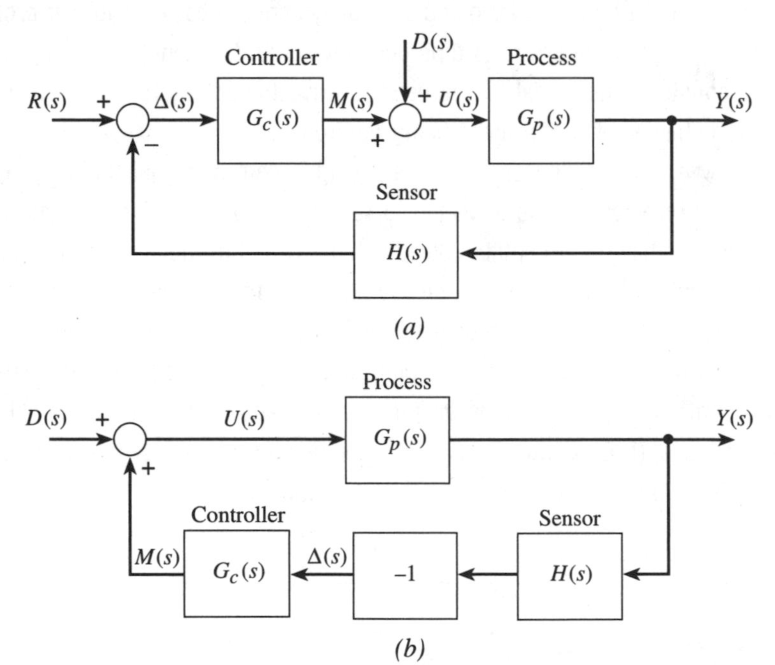

FIGURE 8.1 *Feedback system for PID controller design (a) Block diagram with reference and disturbance inputs (b) Simplified block diagram with disturbance input only*

In general, for the selection of K_P, we take advantage of the computer to perform a sweep of a time simulation for a range of values for K_P and pick the gain that best satisfies the design specifications. In the following example, we will illustrate this approach.

Although we will use only time-domain specifications for the design of PID controllers in this chapter, we should also evaluate the final design using frequency-domain measures such as gain and phase margins and bandwidth to obtain further insight. This evaluation, when not discussed directly in the examples, will be contained in the M-files for the examples available from the Brooks/Cole web site.

EXAMPLE 8.1 ***Proportional Control***

For the feedback control system in Figure 8.1(a) with the process and sensor models

$$G_p(s) = \frac{4}{(2s+1)(0.5s+1)} \quad \text{and} \quad H(s) = \frac{1}{0.05s+1}$$

design a proportional controller $G_c(s) = K_P$ by following the steps given below:

a. Draw the root locus for variations in K_P and use the `rlocfind` command to find K_P^*, the controller gain for which the closed-loop system becomes marginally stable. Also use the `sgrid` command to determine the value of K_P for which the system has a pair of complex closed-loop poles with a damping ratio of $\zeta = 0.8$.

b. Generate a plot that shows the responses to a unit-step reference input for several values of $K_P < K_P^*$. Use the RPI function `tstats` to find the percent overshoot of the step responses. Present the results in a table and determine by trial and error the largest controller-gain value $\hat{K}_P$ that results in a unit-step response having no more than 20% overshoot with respect to the steady-state output value. Determine the gain and phase margins of the feedback system with this gain.

c. Use the gain $\hat{K}_P$ to obtain a single plot showing the responses to unit-step reference and disturbance inputs, taken separately, and determine the values of the steady-state responses.

Solution

a. The MATLAB commands in Script 8.1(a) will generate the root-locus plot shown in Figure 8.2(a). When the command `rlocfind` is used with the cursor placed where the locus crosses the imaginary axis at $s \approx j7$, we find that `kk` yields $K_P^* = 13.95$ and the three corresponding closed-loop poles in `polesCL` are $s = j7.11, -j7.11$, and -22.5. When the `rlocfind` command is used a second time following the command `sgrid(0.8,[ ])`,

PROPORTIONAL-INTEGRAL-DERIVATIVE CONTROL

we find that for $K_P = 0.30$ there will be a pair of complex closed-loop poles having a damping ratio of $\zeta = 0.80$. The three poles are at $s = -1.22 \pm j0.83$ and -20.1. By having the second argument of the `sgrid` command be `[ ]`, which denotes an empty variable, we are telling MATLAB not to draw any lines of constant natural frequency.

The root locus has two branches moving into the right half-plane as K_P is increased, and a third branch approaching infinity along the negative real axis. This asymptotic behavior is due to the fact that the open-loop system has three more poles than zeros.

MATLAB Script

```
% Script 8.1(a): Prop-only design for 2nd-order plant + 1st-order sensor
tauP1 = 2, tauP2 = 0.5, tau_sen = 0.05% process & sensor time constants
Kproc = 4                             % process gain
denG = conv([tauP1 1],[tauP2 1]);     % denominator of forward path
Gp = tf(Kproc,denGp)                  % forward path
H = tf(1,[tau_sen 1])                 % feedback path
GpH = Gp*H;                           % open-loop transfer function
rlocus(GpH)                           % draw root-locus plot with OL model
sgrid(0.8,[ ])                        % lines for damping ratio = 0.8
axis equal; axis([-25 5 -15 15])      % adjust scaling
[kk,polesCL] = rlocfind(GpH)          % gain value for zeta = 0.8
```

CHAPTER 8

b. On the basis of the findings in part (a), we will investigate several step responses for controller gains in the interval $0 < K_P < 13.95$. After some experimentation, we select the range of values $K_P = 0.7, 0.9, \ldots, 1.7$ to illustrate the design. To produce the required plot of the step responses for a sweep of the controller gain, we use a `for` loop to cause `KP` to take on the specified values. For each gain value we build the closed-loop model by multiplying the numerator of $G_p(s)$ by `KP` within the argument list of the `feedback` command. Then the step response is computed and plotted versus time, with the `hold` option in effect for all responses after the first one so the plots will be superimposed on a single set of axes.[1] The `hold` option is released after the last curve has been added to the plot.

[1] The default value of `ishold` is 0, which puts MATLAB in the `hold off` mode. The `hold on` command will set `ishold` to 1, and the `hold off` command will reset it to 0.

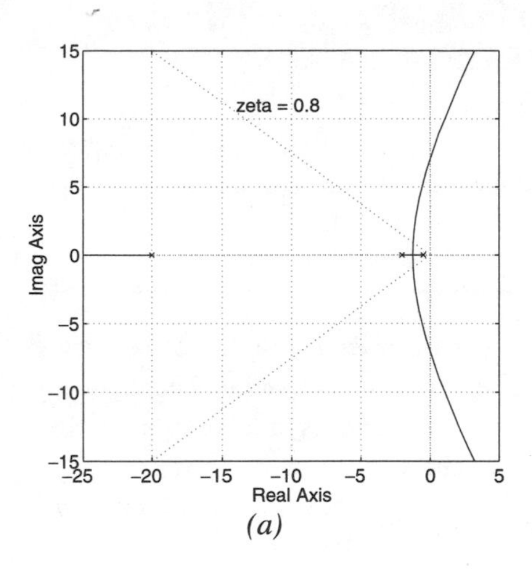

(a)

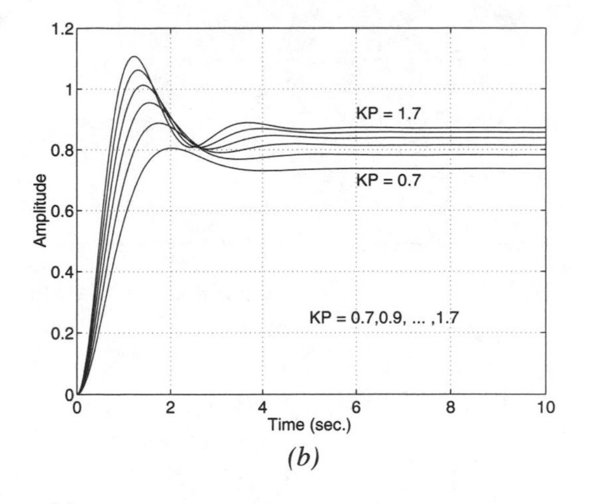

(b)

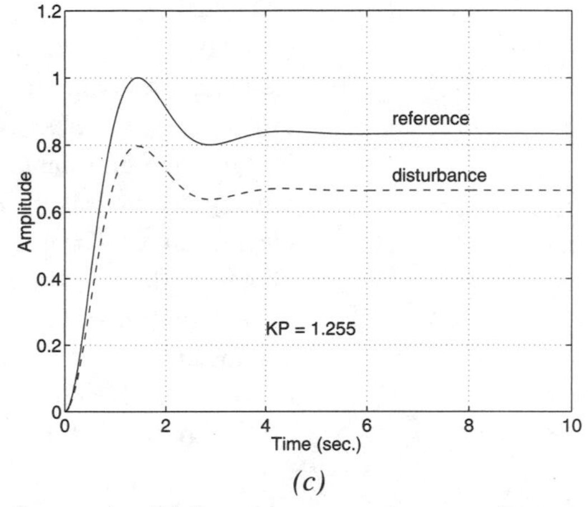

(c)

FIGURE 8.2 *Plots for proportional control (a) Root-locus plot (b) Responses to unit-step reference input for various values of K_P (c) Response to unit steps in the reference and disturbance inputs with $K_P = 1.255$*

PROPORTIONAL-INTEGRAL-DERIVATIVE CONTROL

MATLAB Script

```
% Script 8.1(b):  Step response with proportional gain sweep
%     Use Kproc, denG, numH, denH from Script 8.1(a)
t = [0:0.02:10]';                          % vector of time points
%-- for each value of KP: build CL model, solve,and plot response
for KP = 0.7:0.2:1.7;
  T = feedback(KP*Gp,H,-1);
  ys = step(T,t);
  ref = dcgain(T);                         % steady-state value of y
  [Mo,tp,tr,ts,ess] = tstats(t,ys,ref);    % calc Mo & tp, ignore rest
```

Continues

```
   ymax = ys(find(t==tp));
   yss = ys(length(ys));
   disp([KP ymax tp yss Mo])
   plot(t,ys)
   if ishold ~= 1, hold on, end      % put hold on after first curve
end
hold off                              % turn hold off after last curve
```

Figure 8.2(b) shows the results of the gain sweep. For the lower values of K_P there is a longer rise time, less overshoot, and a larger steady-state error. For the higher values of K_P there is a shorter rise time, more overshoot, and a smaller steady-state error. Clearly, our options for improving the performance of the closed-loop system are rather limited with the proportional controller.

Within the `for` loop of the step response generation, the percent overshoot M_o is computed using the `tstats` function with the step response `y` and the time array `t` as the input arguments. Here the overshoot is defined with respect to the steady-state value of the output y_{ss}, which becomes the third input argument to `tstats`. Because the transients have already decayed at the end of the simulation, y_{ss} is taken to be the last element in `y`. The time to reach the peak value (t_p) is also computed. Table 8.1 shows the numerical results obtained for the range of controller gains specified in Script 8.1(b). By adjusting the values of `KP` in the `for` command to be confined to a smaller interval, or by entering specific values of `KP` manually and executing the instructions that are contained within the `for` loop, we find that $K_P = 1.255$ will result in an overshoot of 20%.

TABLE 8.1 *Performance measures for the step responses generated by the gain sweep in part (b) of Example 8.1*

K_P	y_{max}	t_p (s)	y_{ss}	M_o (%)
0.70	0.8053	2.02	0.7368	9.28
0.90	0.8876	1.74	0.7826	13.42
1.10	0.9553	1.56	0.8148	17.24
1.255	1.0006	1.44	0.8339	20.00
1.30	1.0129	1.42	0.8387	20.77
1.50	1.0630	1.32	0.8571	24.02
1.70	1.1075	1.22	0.8718	27.04

Applying the command `[km,pm,wp,wg] = margin(KP*Gp*H)` with $K_P = 1.255$, we obtain a gain margin of $k_m = 21$ dB. This result indicates that the controller gain that will result in a marginally stable closed-loop system is $K_P^* = 1.255 \times 10^{21/20} = 14.08$. This number is close to the values of 13.95 obtained with the `rlocfind` command. Also, note that the value of the phase-crossover frequency ω_p, at which the gain margin is measured, is 7.14 rad/s. From the root-locus plot in Figure 8.2(a), we see that the upper right root-locus branch intersects the imaginary axis at $\omega \approx 7.0$ rad/s, which agrees quite well with ω_p.

c. With $K_P = 1.255$, we use the `feedback` command to construct the closed-loop system model shown in Figure 8.1(a) and solve for the step response with the `step` command, without plotting it. These operations are implemented in the first half of Script 8.1(c) and result in the column vector `y_ref`. Note that the optional third argument of the `feedback` command has been set to `-1`, its default value, for emphasis.

Then we build the closed-loop model shown in Figure 8.1(b), for which the input is $D(s)$, the sign at the feedback summing junction is positive, and the numerator of the feedback transfer function is negative. Note that the third argument in the `feedback` command is set to `+1` because of the positive feedback shown in the figure. Use of the `step` command produces the column vector `y_dist`. Next we plot the two step responses versus time in Figure 8.2(c). Finally, the steady-state response values are obtained from the last two statements in the script file and are $(y_{\text{ref}})_{\text{ss}} = 0.8339$ and $(y_{\text{dist}})_{\text{ss}} = 0.6645$.

MATLAB Script

```
% Script 8.1(c):  Closed-loop reference & disturbance step responses
tauP1 = 2, tauP2 = 0.5, tau_sen = 0.05   % time constants
Kproc = 4, KP = 1.255                    % process and controller gains
T_ref = feedback(KP*Gp,H,-1)          % build CL systems with reference input
t = [0:0.02:10];                      % define time vector for plots
y_ref = step(T_ref,t);                % CL step response to reference input
T_dist = feedback(Gp,-KP*H,+1)        % build CL system with disturbance input
y_dist = step(T_dist,t);              % CL step response to disturbance input
plot(t,y_ref,t,y_dist,'--')           % plot both responses
y_ref_ss = y_ref(length(t))           % final value for reference input
y_dist_ss = y_dist(length(t))         % final value for disturbance input
```

WHAT IF?

a. Suppose the design specification in part (b) of Example 8.1 is to achieve a rise time of no more than 1 s. Find the smallest value of K_P that will satisfy the specification.

b. Suppose the design specification in part (b) of Example 8.1 is to achieve $y_{ss} \geq 0.9$. Find the smallest value of K_P that will satisfy the specification. ■

PROPORTIONAL-PLUS-INTEGRAL CONTROL

In Example 8.1 we have shown that proportional control can yield well-damped responses to both reference and disturbance step inputs. Although increasing the controller gain will reduce the steady-state errors, it will also increase the overshoot and reduce the damping. Clearly, a more complex controller will be required if we need to have significantly reduced or zero steady-state errors with acceptable overshoot and damping. In this section we will pursue these objectives by introducing an integral term in the controller.

The form of the proportional-plus-integral (PI) controller that we will use is

$$m(t) = K_P\left(\Delta(t) + K_I \int_0^t \Delta(\lambda)\, d\lambda\right)$$

where Δ is the controller input, $m(t)$ is the controller output, as shown in Figure 8.1(a), and λ is a dummy variable of integration. Thus, the transfer function of the controller is $G_c(s) = M(s)/\Delta(s)$, which becomes

$$G_c(s) = K_P\left(1 + \frac{K_I}{s}\right) = K_P\left(\frac{s + K_I}{s}\right) \tag{8.1}$$

Using this controller, we will be able to ensure that a stable feedback control system will achieve zero steady-state errors for step inputs.

Our approach will be to use MATLAB in the same manner as was done in Example 8.1, with sweeps of the controller gain K_P for several values of the integral gain K_I. On the basis of the results we will select a value for K_I to satisfy the design specifications.

EXAMPLE 8.2
PI Control

For the system in Example 8.1, design a PI control law (8.1) according to the following steps:

a. Plot the reference unit-step responses for a sweep of the proportional gain K_P from 0.3 to 1.7 in increments of 0.2 for integral-gain values of $K_I = 0, 0.1, 0.25, 0.5, 1.0$, and 2.0. Select a value $\hat{K}_I$ that will result in a 2% settling time $t_s \leq 6$ s.

b. Plot the root locus for variations in K_P with $K_I = \hat{K}_I$. Then use the `rlocfind` command to calibrate several points on the locus in terms of K_P; find K_P^*, the gain for which the closed-loop system becomes marginally stable.

c. With $K_I = \hat{K}_I$, select the largest K_P such that the overshoot for a reference unit-step input is less than 10%. Then compute and plot the unit-step responses to both reference and disturbance inputs. Comment on the differences between these responses and those in Figure 8.2(c) obtained with a proportional controller.

Solution

a. Script 8.2(a) contains the commands to accomplish the building of the closed-loop model and the generation of the step response plots. We form the denominators of $G(s) = G_c(s)G_p(s)$ and $H(s)$, the forward and feedback transfer functions, respectively, and form the numerator of $H(s)$. The numerator of $G(s)$, which involves the process gain `Kproc` and both of the controller gains, `KP` and `KI`, is not explicitly formed because the proportional gain `KP` is being varied within the `for` loop. So we have used the expression for the numerator of $G(s)$ as the first argument of the `feedback` command.

The response curves for $K_I = 0$ will be the same as those obtained with proportional control, because $G_c(s)$ reduces to K_P in this case. For the other values of K_I, the step responses for $K_I = 0.25, 0.5$, and 1.0 are shown in Figure 8.3. The responses in Figure 8.3(a) for a low integral gain ($K_I = 0.25$) appear to be headed toward $y = 1.0$, but the transients are dying out slowly, resulting in a large settling time t_s. Increasing K_I to 0.5 causes the transients to be reduced to virtually zero for all $t > 6$ s, for all values of K_P used. For a high integral gain ($K_I = 1.0$), Figure 8.3(c) shows that the responses no longer have sufficient damping, resulting in a longer settling time. From this analysis, we conclude that the responses for $K_I = 0.5$ will satisfy the constraint of 2% settling time $t_s \leq 6$ s. Accurate values of t_s can be found using the `tstats` command. As a result, we select $K_I = 0.5$.

MATLAB Script

```
% Script 8.2(a): Sweep of KP for selected values of KI
tauP1 = 2, tauP2 = 0.5, Kproc = 4          % process time constants & gain
tau_sen = 0.05                             % sensor time constant
%----- forward transfer function for reference input is G = Gc * Gp
denG = conv([tauP1 1],[tauP2 1]);
Gp = tf(Kproc,denGp)
H = tf(1,[tau_sen 1])
GpH = Gp*H;
%----- compute CL step responses for range of KP values
t = [0:0.02:10]';
KI = input('enter integral gain KI ==>');    % user specifies KI
```

Continues

```
for KP = 0.3:0.2:1.7                          % repeat for different KP
  Kc = tf(KP*[1 KI],[1 0]);                   % numerator depends on KP
  T = feedback(Kc*Gp,H,-1);                   % closed-loop system
  ys = step(T,t);                             % closed-loop step response
  [Mo,tp,tr,ts,ess] = tstats(t,ys,1)          % compute settling time
  plot(t,ys)
  if ishold ~= 1, hold on, end             % plots for all KP on same axis
end
hold off
```

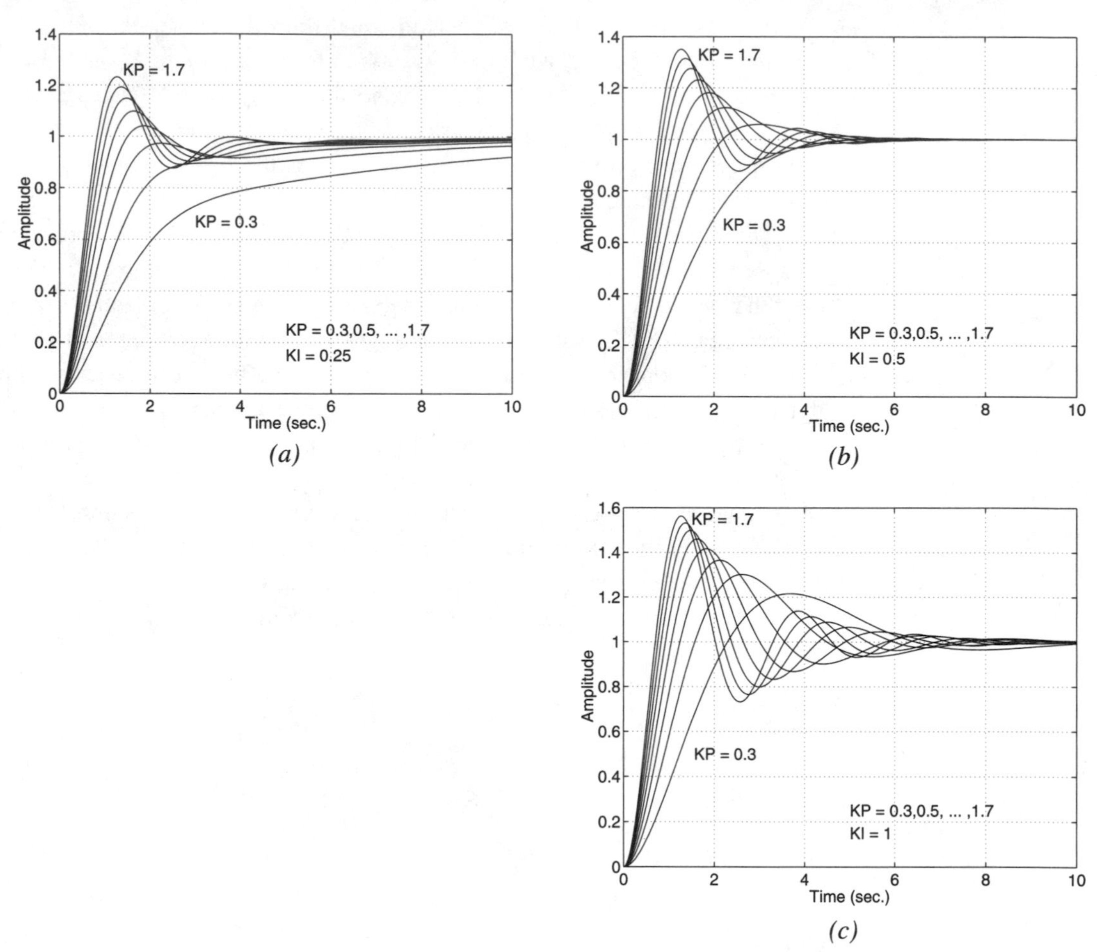

FIGURE 8.3 *Results of proportional-gain sweep with PI controller (a) $K_I = 0.25$ (b) $K_I = 0.5$ (c) $K_I = 1.0$*

b. Using Script 8.2(b) for $K_I = 0.5$, the required root locus is shown in Figure 8.4. Note that the controller transfer function $G_c(s)$ has a pole at

$s = 0$ and a zero at $s = -K_I = -0.5$, resulting in the controller zero canceling the process pole at $s = -0.5$.

By repeated use of the `rlocfind` command, we have determined the values of K_P corresponding to five points on the upper complex branch of the locus. We see that for $K_P = 0.24$ the closed-loop system will have a double pole at $s = -1.0$. The numerical value of the pole can be displayed in the MATLAB command window by using the `rlocfind` command, where `kk` contains the required proportional gain and the array `polesCL` gives the corresponding closed-loop poles.

To see the effect of the change in going from proportional control to PI control, compare the root-locus plots in Figures 8.2(a) and 8.4. Observe that the net effect of the integral term with $K_I = 0.5$ is to replace the open-loop process pole at $s = -0.5$ with a new pole at $s = 0$. This change has shifted the complex portion of the locus for the PI controller slightly to the right, thereby making the complex branches cross into the right half-plane at a lower value of ω (about 6.3 rad/s versus 7.1 rad/s for proportional control). Also the point at which the large-gain asymptotes intersect has been shifted slightly to the right ($\sigma_0 = -7.50$ for proportional control versus -7.333 for PI control). There is also a third branch of the locus, starting at the open-loop pole at $s = -20$ and going to the left along the negative real axis. The transients associated with this pole always decay much faster than those with the other poles, thus having very little effect on the responses.

MATLAB Script

```
% Script 8.2(b):  Root-locus plot for PI control
% Use KP, Gp, and H from Script 8.2(a)
KI = 0.5
Kc = tf(KP*[1 KI],[1 0]);
rlocus(Kc*Gp*H)                    % draw root-locus plot for OL model
axis([-8 2 -2 8])                  % adjust region plotted
[kk,polesCL] = rlocfind(Kc*Gp*H)   % calibrate locus with gains
```

c. On the basis of the responses shown in Figure 8.3(b), we can estimate that a value of K_P midway between 0.5 and 0.7 will result in approximately 10% overshoot. Hence, we will settle on $K_P = 0.6$. The reference and disturbance step responses can be computed with a M-file that is almost identical to Script 8.1(c). The only changes required are: (i) the controller transfer function has the numerator `KP*[1 KI]` and its denominator is `[1 0]`, and (ii) the user must be allowed to specify the integral gain `KI` in addition to `KP`. The results are shown in Figure 8.5 and illustrate that the steady-state errors for both inputs go to zero. For the reference step response,

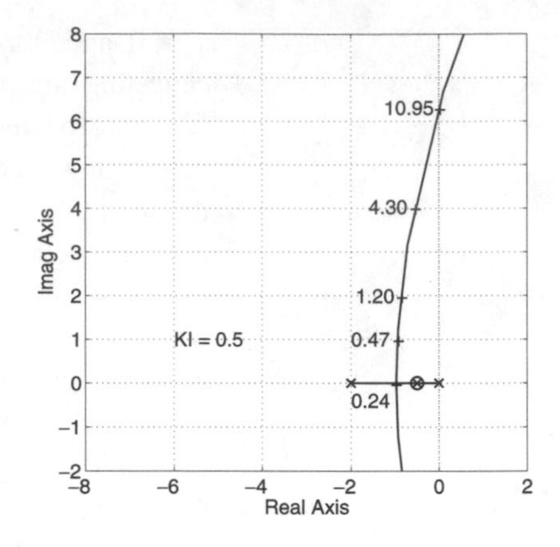

FIGURE 8.4 *Root-locus plot for PI control with gain calibrations added*

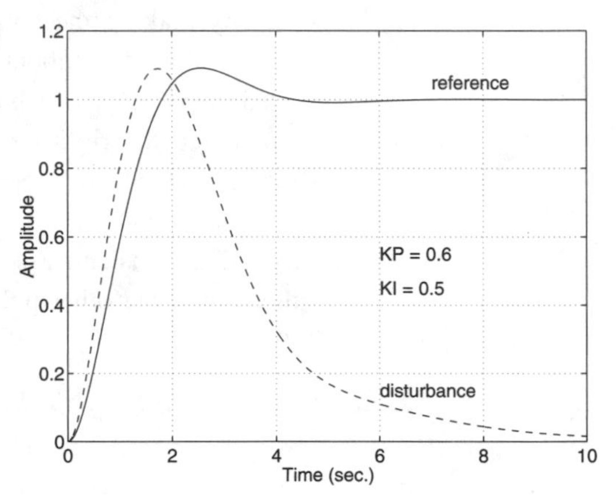

FIGURE 8.5 *Reference and disturbance step responses with PI controller gains of $K_P = 0.6$ and $K_I = 0.5$*

CHAPTER 8

there is approximately 10% overshoot and a 2% settling time of less than 4 s, satisfying the design specifications. The response to the disturbance input reaches a somewhat higher value than with the proportional controller (1.1 versus 0.8), but it settles out at zero in just over 10 seconds, whereas with proportional control the steady-state error is approximately 0.65. Thus we have graphic evidence, at least for the process and sensor models used, of the ability of the PI controller to eliminate steady-state errors to both reference and disturbance step inputs. Essentially, the integral term in the PI controller has increased the system type from 0 to 1. It is well known that a type-1 system will exhibit zero steady-state error to a step input.

WHAT IF?

a. If $K_I = 0.25$ or 1.0 is used in the PI controller, there will no longer be a pole-zero cancellation. Draw root-locus plots for these values of K_I

and examine closely the change of the behavior of the root locus near the origin of the s-plane. Relate your observations to the step responses shown in Figures 8.3(a) and (c).

b. Suppose the reference and disturbance inputs are ramp functions. What sort of steady-state errors would you expect? Copy the file `fig8_5.m` that produced Figure 8.5 and modify it to produce the responses of this closed-loop system to ramp inputs with unity slopes, taken one at a time. ■

PROPORTIONAL-INTEGRAL-DERIVATIVE CONTROL

With a PI controller, we would expect a control system to satisfy design specifications of overshoot, settling time, and zero steady-state error. However, if a faster rise time t_r without increased overshoot is desired, then a derivative term may have to be included in the controller. With the derivative term included, the proportional-integral-derivative (PID) control law is

$$m(t) = K_P\left(\Delta(t) + K_I \int_0^t \Delta(\lambda)d\lambda + K_D \frac{d\Delta}{dt}\right)$$

which leads to the transfer function

$$G_c(s) = K_P\left(\frac{K_D s^2 + s + K_I}{s}\right) \tag{8.2}$$

We see that $G_c(s)$ still has the pole at $s = 0$, but now there are two zeros which are located at $s = (-1 \pm \sqrt{1 - 4K_DK_I})/2K_D$. Although it is not necessary, we will restrict our attention to PID controllers whose transfer functions have real zeros. This condition will be satisfied if $1 - 4K_DK_I > 0$, which is equivalent to $K_DK_I < 0.25$.

We now extend the designs of Examples 8.1 and 8.2 by incorporating a derivative term in the controller. The same gain-sweep approach will be used. However, the gain sweep for a PID design will be more involved as there are more interactions between the proportional, derivative, and integral control actions. In general, the integral gain K_I can still be set as if designing a PI controller. However, the proportional and derivative gains K_P and K_D need to be tuned together with two performance measures, namely, overshoot and rise time, to be satisfied. In addition, it is likely that there will be several combinations of K_P and K_D that can satisfy the specifications.

EXAMPLE 8.3
PID Control

For the feedback control system in Example 8.1, design a PID controller such that the unit-step reference response satisfies the specifications of (i) overshoot less than 10%, (ii) rise time $t_r \leq 0.1$ s, and (iii) 2% settling time $t_{s2} \leq 6$ s by following the procedure given below.

In the PI design of Example 8.2, we have determined that the integral gain $K_I = 0.5$ will satisfy $t_{s2} \leq 6$ s. As a result, we fix K_I at that value and adjust K_P and K_D using the following steps:

a. With $K_I = 0.5$, plot the reference unit-step responses for a sweep of the proportional gain K_P from 1 to 10 in increments of 1 for derivative-gain values of $K_D = 0, 0.1, \ldots, 0.5$. On the basis of the results, select the smallest value of K_D, denoted by $\hat{K}_D$, that will allow the overshoot and rise-time specifications to be satisfied without violating the settling-time specification.

b. Plot the root locus for variations of K_P with $K_I = 0.5$ and $K_D = \hat{K}_D$. Then use the `rlocfind` command to calibrate several points on the locus in terms of K_P. Find K_P^*, the gain for which the closed-loop system becomes marginally stable.

c. With $K_I = 0.5$ and $K_D = \hat{K}_D$, select a value of K_P, denoted by $\hat{K}_P$, such that all three specifications are satisfied. Compute and plot the unit-step responses to both reference and disturbance inputs. Comment on the differences between these responses and those in Figure 8.2 (c) obtained with a proportional controller and those in Figure 8.5 for PI control.

Solution

The M-files required for the PID controller design can be readily obtained by editing the files that solve Example 8.2 to incorporate the PID controller and will not be displayed here.

a. We show in Figure 8.6 the step-response plots with $K_D = 0.3, 0.4$, and 0.5 for a sweep of proportional gain K_P. We observe that in Figure 8.6(a) the choice of $K_D = 0.3$ results in either an overshoot M_o greater than 10% or a rise time $t_r > 0.1$ s. However, Figures 8.6(b) and (c) show that these specifications can be satisfied for $K_D = 0.4$ and 0.5. We select the smaller value because it results in responses that avoid some of the undershooting of the final value that is present for $K_D = 0.5$.

b. Figure 8.7 presents two views of the root locus with $K_I = 0.5$ and $K_D = 0.4$. Figure 8.7(a) covers a large enough area in the s-plane to include all four branches (there are four open-loop poles). We see that two branches approach infinity as the proportional gain K_P becomes large, and these branches remain in the left half-plane, regardless of how large the gain is. The details of the two branches that are near the origin are shown in Figure 8.7(b), which was generated by giving the command `axis([-2.5`

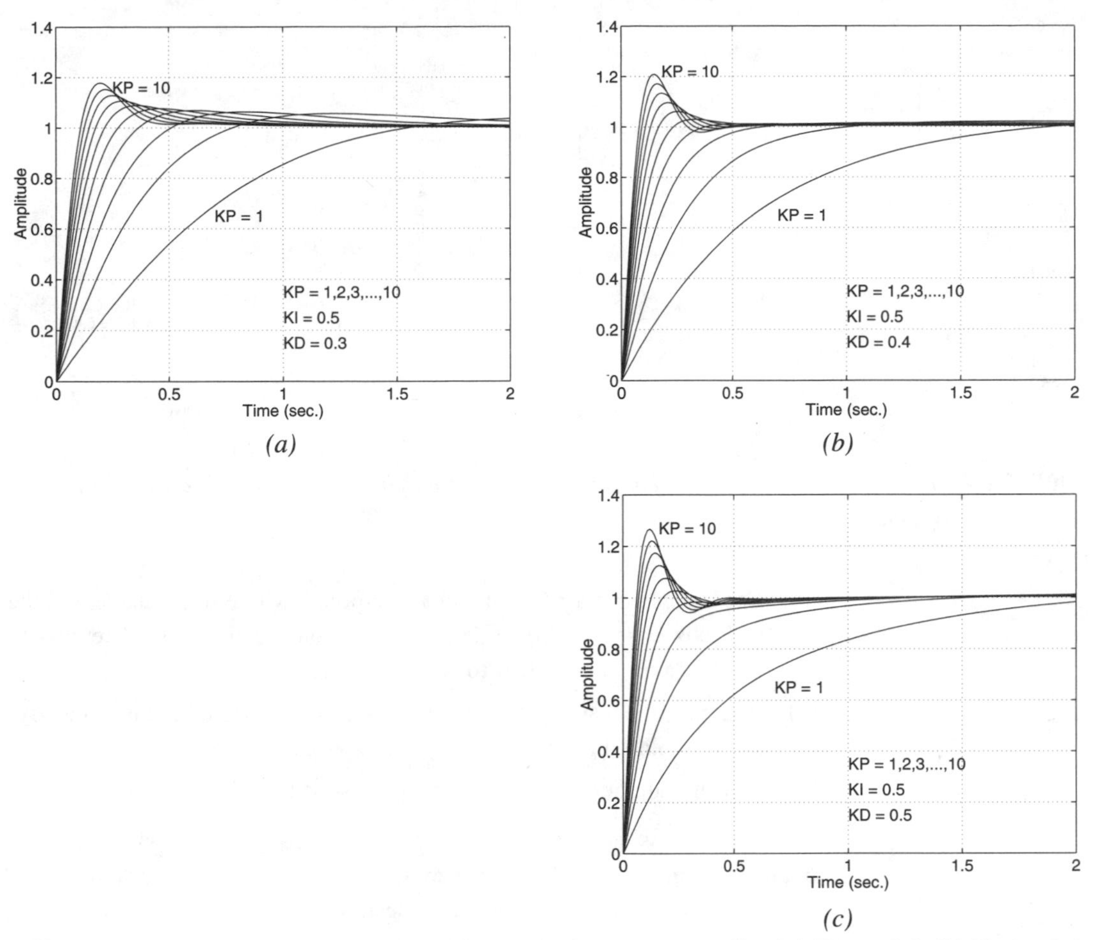

FIGURE 8.6 *Results of proportional gain sweep with a PID controller (a)* $K_D = 0.3$ *(b)* $K_D = 0.4$ *(c)* $K_D = 0.5$

`0.5 -1.5 1.5])` after the first plot had been drawn. Now we can see that the two zeros of the PID controller are at $s = -0.7$ and -1.8 and that each zero has one branch of the locus terminating on it.

Should MATLAb fail to produce root-locus plots with the circular portions appearing as shown in Figure 8.7, it is possible to use the `rlocus` command with its optional gain input argument. To establish an acceptable gain vector `k`, the following steps can be taken:

(i). Draw the locus with the gain vector as an output argument,

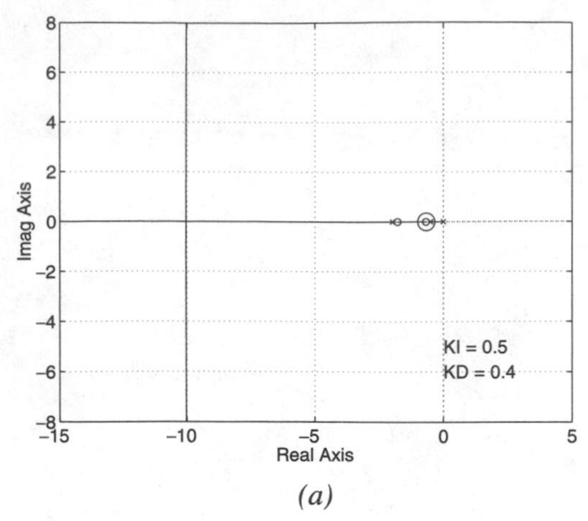

(a)

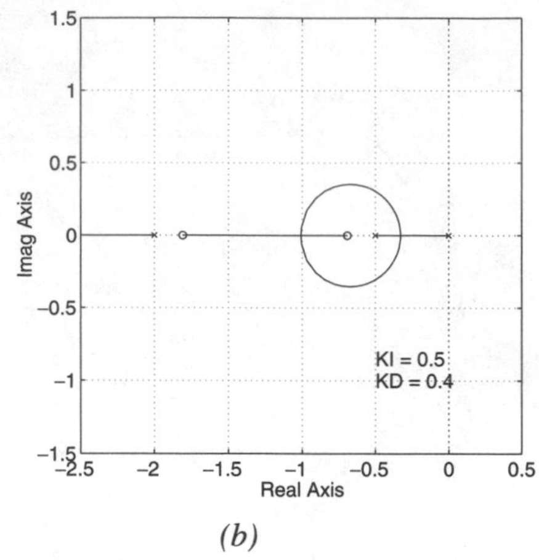

(b)

FIGURE 8.7 *Root-locus plots for PID control (a) Large area showing all four branches (b) Small area showing details near the origin*

(ii). Use the `rlocfind` command at the points where the locus leaves the real axis ($s = -0.35$) to determine `k1` and again where it returns to the real axis ($s = -1.0$) to determine `k2`,

(iii). Augment the gain vector by including a number of logarithmically-spaced points in the interval `[k1,k2]`, and

(iv). Sort the augmented set of gains and use it to redraw the locus.

These steps are illustrated in the file `fig8_7b.m`, which can be obtained from the Brooks/Cole web site.

To obtain a rapid response, a designer must pay attention to the locations of these zeros because there will always be a closed-loop pole located on each of these branches. In particular, for a sufficiently large value of K_P, there will be a closed-loop pole at $s = -1$ or to the right of it. This means that the closed-loop system will always have a mode function whose time constant is no less than 1 s. If the combination of inputs and initial conditions is such as to excite this mode to a significant degree, the response will have a transient that will take approximately four seconds to decay to less than 2% of its original value. This situation is caused by the presence of the integral term in the controller, and it is the price one must pay to achieve a zero steady-state error. Note that the design of the PI controller in Example 8.2 results in a zero-pole cancellation that avoids this phenomenon.

c. Examining Figure 8.6(b), we select $K_P = 7$, $K_I = 0.5$, and $K_D = 0.4$ as this design will satisfy all the specifications. The responses of this design to step-function reference and disturbance inputs are shown in Figure 8.8.

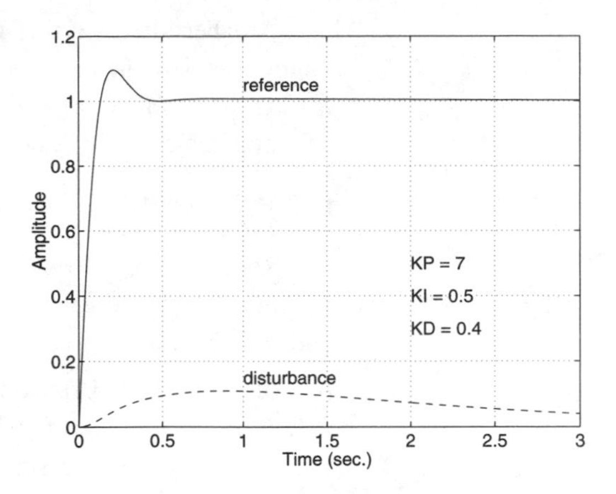

FIGURE 8.8 *Reference and disturbance step responses for PID control with final controller gains*

A comparison of the reference-input responses for the three cases shows a marked improvement for the response with PID control over both of the others. The response with proportional control is deficient in both speed of response and steady-state error. The response with PI control has zero steady-state error, but its rise time is much slower than that with PID control.

Likewise, the disturbance-input response obtained with PID control shows a peak of about 0.1 and zero steady-state error, which is a significant improvement over the other two cases. The disturbance response with proportional control has a peak of 0.8 and a steady-state error of over 0.6. The response for PI control has zero steady-state error, but a peak error of 1.1. Clearly the PID controller has provided greatly improved performance over the other two, at very little increase in complexity.

Comment: One important consideration that has not yet been mentioned is the fact that the PID controller is not physically realizable. This characteristic is reflected in the fact that its transfer function has more zeros than poles. Even if it were realizable, one should not attempt to do perfect differentiation because high-frequency noise will be accentuated. Also, perfect integration is difficult to obtain because of amplifier drift.

In practice, the derivative term in the controller (8.2) is typically implemented as a high-pass filter such that the transfer function for the PD controller is

$$G_c(s) = K_P\left[1 + K_D\left(\frac{s}{Ts + 1}\right)\right] = K_P\left[\frac{(K_D + T)s + 1}{Ts + 1}\right] \tag{8.3}$$

where the time constant T is selected so as to satisfy the relationship $T << K_D$.

Another alternative is to use lead and lag transfer functions, along with gain, to implement an approximation to the PID control law. Such an approach avoids the practical difficulties mentioned above and can be expected to result in comparable performance of the closed-loop system.

WHAT IF?

a. Repeat the root-locus plots of Part (b) of Example 8.3 for values of $K_D =$ 0.3 and 0.5, which correspond to the step responses shown in Figures 8.6(a) and (c). Examine the change in the root locus near the origin of the s-plane.

b. Investigate the PID controller for values of K_I and K_D other than 0.5 and 0.4, respectively. See if you can obtain any improvements in the responses, such as faster rise and settling times.

c. Design a proportional-plus-derivative (PD) controller for the feedback system considered in these examples. The control law is that of the PID controller (8.2), with the integral gain K_I set to zero. You should be able to obtain a rapid response, but there will be steady-state errors comparable to those obtained with proportional control.

d. Suppose the reference and disturbance inputs are ramp functions. What sort of steady-state errors would you expect? Copy the file `fig8_8b.m` that produced Figure 8.8 and modify it to obtain the responses of this closed-loop system to ramp inputs with unity slopes, taken one at a time. ■

EXPLORATION

E8.1 Type-0 plant. Consider the fifth-order plant having the transfer function

$$G_p(s) = \frac{500(s+10)}{(s+2)(s+5)(s^2+6s+25)(s+20)} \quad \text{and} \quad H(s) = 1$$

Design a PID controller that will result in a reference step response that satisfies as many of the following specifications as possible (hopefully all of them):

a. no more than 10% overshoot
b. 10–90% rise time no more than 0.6 s
c. time to peak no more than 1.5 s
d. 2% settling time no more than 2.8 s
e. zero steady-state error

Your MATLAB program should use the RPI function `tstats` to compute the above performance measures for the step response of your final design.

Also, implement the derivative term of the controller as a high-pass filter with time constant $T << K_D$, as discussed in the comments following Example 8.3 and shown in (8.3). Use a root-locus plot and appropriate gain sweeps to help obtain a satisfactory set of controller gains.

E8.2 Type-1 plant. A type-1 plant $G_p(s)$ and a sensor $H(s)$ have the transfer functions

$$G_p(s) = \frac{12{,}000}{s(s+10)(s+30)(s+40)} \quad \text{and} \quad H(s) = \frac{1}{0.02s+1}$$

Design a PD controller that will provide a step response with no more than 10% overshoot and a 2% settling time of no more than 1 s. Verify that the steady-state error to the step input will be zero but that there will be a nonzero steady-state error to a ramp input. Calculate the velocity constant of your design and relate it to the steady-state error.

COMPREHENSIVE PROBLEMS

CP8.1 Electric power generation system. Use the gain-sweep approach to design the controllers outlined below. In this design, we assume that some supplementary damping has been added to the electromechanical mode, so that the regulator design would not be affected by the stability of this mode. This is done by changing the natural damping term in the model and has resulted in the model file `epow2.m`, which should be used to obtain the process model for the design.

a. P-controller design: Design a proportional controller $G_c(s) = K_P$ for the voltage control of the electric power system shown in Figure A.2. For a unit-step reference input, the overshoot should not exceed 15%.
b. PI-controller design: Design a PI controller (8.1) for the voltage control of the electric power system shown in Figure A.2. For a unit-step reference input, the overshoot should be no more than 15%, the rise time no more than 0.5 s, and the 2% settling time no more than 5 s.

CP8.2 Satellite with TF model. Obtain the model of the open-loop satellite by running the file `sat.m` and selecting the transfer function from the motor torque to the pointing angle. This form of the satellite model is subject to the restriction that the initial wheel speed is zero. Because of this restriction, you can design a PD controller that will have zero steady-state error to a step change in the desired pointing angle.

First, draw a root-locus plot to convince yourself that the proportional gain must be negative in order to have a stable closed-loop system. Can you explain why? For a derivative gain in the range of $5 \le K_D \le 50$, use the `rlocfind` and `sgrid` commands to obtain an approximate value of the proportional gain that will result in dominant closed-loop poles having a damping ratio in the interval $1.0 \leq \zeta \leq 0.7$. To have a controller transfer function that is proper (no more zeros than poles), include a pole with the derivative term, as illustrated in (8.3). A reasonable choice for the time constant associated with the controller pole is to make $T = K_D/20$.

Determine a set of controller gains (K_P and K_D) that will result in a pointing-angle step response with no more than 10% overshoot and a 2% settling time of no more than 100 s. This should not be difficult to do because no restriction has been placed on the speed of the reaction wheel. In fact, we don't even see the wheel speed with the transfer-function model being used for the design.

CP8.3 Satellite with SS model. Use the file `sat.m` to obtain the four matrices of the state-space satellite model where the input is the torque applied to the reaction wheel. Note that this model has two outputs [the pointing angle, in degrees, and the wheel speed, in revolutions per minute (rpm)]. Include a PID controller and use the command `lsim` with the initial condition on the state vector to simulate the step response with a nonzero initial wheel speed (say, 2000 rpm). With the integral gain $K_I = 0$ and the same values for K_P and K_D as in the previous problem, you should find that you get the same response as the TF model gave when the initial wheel speed is zero, but the steady-state pointing error will no longer be zero when the initial wheel speed is nonzero. To get a zero steady-state pointing error for any initial wheel speed, it is necessary to use an integral term in the controller by having $K_I > 0$.

Use the commands `subplot(2,1,1)` and `subplot(2,1,2)` in the appropriate places so the actual and desired pointing angles are shown in the upper half of the graphics window and the wheel speed is shown in a second plot directly below it. To get a set of initial conditions on all of the state variables that will start the simulation close to equilibrium, obtain the zero-input response with the nonzero initial wheel speed and take the final state vector after the transients have died out as the initial conditions of a run with a step change in the desired pointing angle. To see the effects of the step input, delay it by 200 seconds so any remaining transients due to the initial conditions will settle out.

Obtain gains for a PID controller that will result in the following specifications for a 1° step change in the desired pointing angle when the initial wheel speed is 2000 rpm:

a. no more than 25% overshoot
b. 2% settling time of no more than 100 s
c. maximum change in wheel speed of 1500 rpm

SUMMARY

We have discussed the design of PID controllers to satisfy time-domain performance measures discussed in Chapter 7. Because there are no exact relationships between the PID gain parameters and the time-domain performance measures for high-order systems, a gain-sweep approach based on time simulation is used to select controller gains to satisfy the performance measures. We have also used root-locus plots to improve our understanding of the effects of the controller gains on the step responses. In Chapter 9 we will consider the design of lead-lag controllers using frequency-response methods and as an approximation to a PID design.

MATLAB FUNCTIONS USED

Function	*Purpose and Use*	*Toolbox*
*	Given two LTI objects, the * operator forms their series connection.	Control System
axis	**axis([xmin xmax ymin ymax])** specifices the plotting area. **axis equal** forces uniform scaling for the real and imaginary axes.	MATLAB
conv	Given two row vectors containing the coefficients of two polynomials, **conv** returns a row vector containing the coefficients of the product of the two polynomials.	MATLAB
feedback	Given the models of two systems in TF form, **feedback** returns the model of the closed-loop system, where negative feedback is assumed. An optional third argument can be used to handle the positive feedback case.	Control System
hold	When set "on," **hold** draws subsequent plots on the current set of axes.	MATLAB
margin	Given a model in TF or SS form, **margin** returns the gain and phase margins and the crossover frequencies. When the output variables are omitted, it generates a Bode plot with the margins and crossover frequencies indicated on the plot.	Control System

rlocfind	Given a TF or SS model of an open-loop system, **rlocfind** allows the user to select any point on the locus with the mouse and returns the value of the loop gain that will make that point a closed-loop pole. It also returns the values of all the closed-loop poles for that gain value.	Control System
rlocus	Given a TF or SS model of an open-loop system, **rlocus** produces a root-locus plot that shows the locations of the closed-loop poles in the s-plane as the loop gain varies from 0 to infinity.	Control System
sgrid	When viewing either a root-locus plot or a pole-zero map, **sgrid** draws contours of constant damping ration (ζ) and natural frequency (ω_n).	Control System
step	Given a TF or SS model of a continuous system, **step** returns the response to a unit step function input.	Control System
subplot	**subplot** allows the plotting window to be divided into multiple plotting areas.	MATLAB
tf	Given numerator and denominator polynomials, **tf** creates the system model as a TF object. The command also converts zero-pole-gain or state-space models to TF form.	Control System
tstats	Given a step response, **tstats** finds the percent overshoot, peak time, rise time, settling time, and steady-state error.	RPI function

CHAPTER 9

Frequency-Response Design

PREVIEW

In the previous chapter, we presented several examples in which proportional, proportional-integral (PI), and proportional-integral-derivative (PID) compensators were designed. However, the integral and derivative operations are not strictly realizable in practice, and one usually implements the controller with lead and lag elements that were introduced in Chapter 3. In this chapter we use MATLAB to design lead-lag controllers using frequency-response specifications and design methods. As there is no standard notation or definition of the lead and lag transfer functions, the reader should take care to be sure just how the transfer functions and parameters that are being used have been defined. The discussion and examples of this chapter are based on the treatments of Franklin, Powell, and Emami-Naeini (1994) and Kuo (1995).

LAG CONTROLLER DESIGN

When designing a lag compensator for feedback systems having the structure shown in Figure 9.1, which is the same as Figures 7.1(a) and 8.1(a), we want to attain at least a specified phase margin (for dynamic response) and low-frequency gain (for steady-state error to a step input). When using frequency-response methods, there are two ways in which the design can be approached. One method is to achieve a satisfactory phase margin and use the lag to increase the low-frequency gain. The other approach is to start with the gain set at a value that will meet the steady-state error specification and then to use the lag to reduce the mid- and high-frequency magnitude so the phase-margin requirement can be met. We will summarize these two methods and then illustrate their use by applying them to the plant and sensor that were used in the illustrative examples in Chapter 8.

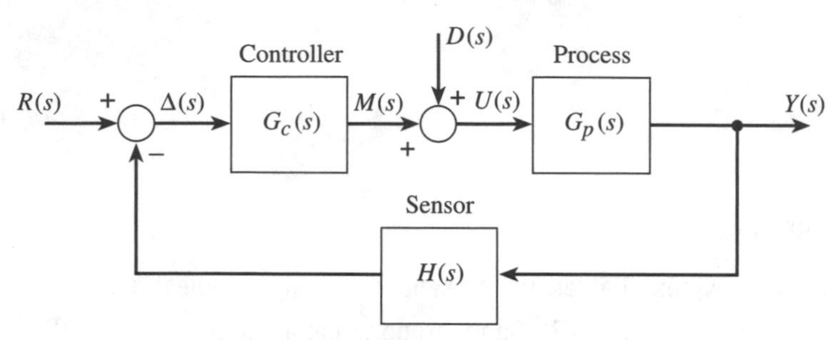

FIGURE 9.1 *Feedback system for controller design*

METHOD 1—RAISING LOW-FREQUENCY MAGNITUDE

Franklin, Powell, and Emami-Naeini (1994) present a method for the design of lag compensators that attempts to satisfy phase-margin and low-frequency-gain specifications and starts at the high end of the frequency spectrum. We paraphrase their lag compensation design procedure below. The controller transfer function is written as

$$G_c(s) = K_{\text{lag}}\left(\frac{s - z_{\text{lag}}}{s - z_{\text{lag}}/\alpha_{\text{lag}}}\right) \tag{9.1}$$

where $\alpha_{\text{lag}} > 1$. Given desired values for the phase margin and for the low-frequency gain of the open-loop system (to ensure attaining a desired steady-state error to a constant reference input), we must determine values for the gain K_{lag}, for the zero-pole ratio α_{lag}, and for the controller zero z_{lag}. The magnitude of z_{lag} is also known as the *corner frequency* of the zero. The steps that we will take to accomplish this are:

1. With a proportional-only controller, find the gain K_{lag} such that the phase-margin specification is satisfied, without being concerned with the low-frequency gain. Having done this, we designate the gain-crossover frequency as ω_c and denote the low-frequency gain attained with this controller as A. To be on the safe side, we may

adjust the value of K_{lag} to obtain a bit more than the required phase margin (say, 5° to 10°).

2. We compute the value of α_{lag} as the ratio of the desired low-frequency gain to A, the gain attained with the proportional-only control in the previous step.
3. We select the corner frequency of the lag to be between one decade and one octave below ω_c, the magnitude-crossover frequency found in step 1. Hence, we look for $\omega_c/10 \le -z_{\text{lag}} \le \omega_c/2$. The compensator pole will be $s = z_{\text{lag}}/\alpha_{\text{lag}}$. Then we draw a Bode plot of the open-loop system with the lag controller in series with the plant and the sensor or use the `margin` command to determine the actual phase margin. If necessary, we repeat this process for different values of z_{lag} until we are able to meet the phase-margin specification. At this point the initial design is complete and we plot the responses of the system to reference and disturbance step inputs in order to assess the behavior in the time domain.

Depending on the step responses, we may want to iterate on one or more of the three design parameters to attain a more satisfactory response. For example, the reference step response may have more overshoot than desired, or perhaps it has a slowly decaying transient. To deal with these situations, we can adjust K_{lag} to increase the phase margin and increase the magnitude of z_{lag} to make the transient decay faster.

In the following example we will apply these steps to the design of a lag controller for the plant and sensor used in the examples throughout Chapter 8. The system is type-0, so the steady-state error is $e_{\text{ss}} = 1/(1 + K_p)$ where K_p is the position constant, given by (7.3). Because we will not be able to attain zero steady-state error to a step reference input as we did with the PI controller in Example 8.1, we will settle for a specification of 2% steady-state error. This means that the low-frequency gain of the open-loop system must be at least $(1/0.02) - 1 = 49$. Also, by making a Bode plot of the open-loop frequency response of the final design in Example 8.1, we find that its phase margin is 60.4°. Hence, we will try to attain a phase margin of at least 60° with the lag controller.

EXAMPLE 9.1
Lag Controller Design, Method 1

Using the frequency-response design algorithm described above, find a lag compensator that will result in a phase margin of approximately 60° and an open-loop low-frequency gain of 49 for the plant and sensor whose transfer functions are

$$G_p(s) = \frac{4}{(2s + 1)(0.5s + 1)} \quad \text{and} \quad H(s) = \frac{1}{0.05s + 1} \tag{9.2}$$

Solution We will present the design in two parts. First, we determine values of $K_{\rm lag}$ and $\alpha_{\rm lag}$ that will provide the required low-frequency gain and allow for meeting the phase-margin specification when the lag zero is selected. In the second part, we try one or more values for the lag zero and see what we get for the phase margin.

a. The commands in Script 9.1(a) build a series interconnection of the open-loop plant $G_p(s)$ and the sensor $H(s)$. Then the magnitude and phase of the open-loop frequency response are computed at 100 logarithmically spaced frequencies in the interval $0.1 \le \omega \le 10$. The output variables `mag` and `ph` are three-dimensional arrays (to allow for multi-input/multi-output systems). Because we are dealing with single-input/single-output systems, we use the `reshape` command to convert them into column vectors.

Allowing a 10° safety margin, we want to find the gain $K_{\rm lag}$ that will result in $60 + 10 = 70°$ of phase margin. This can be done by using MATLAB to search the phase-angle results of the `bode` command and find the index of the entry for which the phase angle is closest to $70 - 180 = -110°$. Then we set $K_{\rm lag}$ equal to the reciprocal of the corresponding magnitude value. To assist in this process, we use the `find` command, as shown in the script, to extract the index, phase angle, magnitude, and frequency of those entries having phase angles in the interval $[-120°, -100°]$. The result is shown in Table 9.1, from which we can see that the index ii = 58 corresponds to a phase of $-110.0°$, a magnitude of 1.083, and a frequency of 1.42 rad/s. Thus, by making $K_{\rm lag} = 1/1.083 = 0.924$, we will have the desired phase margin, and the gain-crossover frequency will be $\omega_c = 1.42$ rad/s.

MATLAB Script

```
% Script 9.1(a) Determination of Klag, alpha, and w_c
tauP1 = 2; tauP2 = 0.5; tau_sen = 0.05; Kproc = 4; % parameters
Gp = tf(Kproc,conv([tauP1 1],[tauP2 1]);          % plant Gp(s)
H = tf(1,[tau_sen 1]);                             % sensor H(s)
GpH = Gp*H;                         % OL transfer function is plant * sensor
%---------- open-loop frequency response
w = logspace(-1,1,100)';       % use 100 points for better resolution
[mag,ph] = bode(GpH,w);        % computes magnitude & phase as 1:1:100 arrays
%-------- convert magnitude & phase to 100-element column vectors
mag = reshape(mag,100,1);
ph = reshape(ph,100,1);
%-------- display table of index, phase, mag, & freq values
%          by selecting -120 <= phase <= -100 deg
for i = find((ph <= -100)& (ph >= -120)),
```

Continues

```
  disp([ i  ph(i)  mag(i)  w(i)])
end
%-------- interpolate to get magnitude for 70 deg phase margin
mag110 = interp1(ph,mag,-110)
Klag = 1/mag110                         % set gain for selected phase
lfg = Klag*dcgain(GpH);                 % lfg for plant + gain
Alag = 49/lfg    % lag alpha supplies rest of req'd low-freq gain
```

TABLE 9.1 *Selected frequency-response values for the plant and sensor in Example 9.1*

index	*phase (deg)*	*magnitude*	*frequency (rad/s)*
54	−100.8	1.346	1.18
55	−103.1	1.277	1.23
56	−105.4	1.211	1.29
57	−107.7	1.146	1.35
58	−110.0	1.083	1.42
59	−112.2	1.022	1.48
60	−114.5	0.963	1.56
61	−116.8	0.907	1.63
62	−119.0	0.852	1.71

A more direct way of finding the magnitude corresponding to a phase angle of $-110°$ is to use MATLAB's `interp1` command to do an interpolation with the vectors `mag` and `ph`. If we think of the magnitude as a function of the phase angle, we want to know what magnitude value corresponds to the phase angle $-110°$. Rather than building Table 9.1 and doing an approximate interpolation visually, we can have MATLAB obtain K_{lag} by entering the command `mag110 = interp1(ph,mag, -110)` followed by `Klag = 1/mag110`. Doing the calculation this way gives $K_{\text{lag}} = 1/1.082 = 0.924$, which agrees with our previous result after rounding to three decimal places. To get an accurate value for the frequency ω_c at which the phase angle is $-110°$, we can reuse the command `interp1` as `wc = interp1(ph,w,-110)`. The result is $\omega_c = 1.42$ rad/s.

The magnitude and phase plots for the plant and sensor alone are shown in Figure 9.2, from which we can see the graphical significance of these calculations. Horizontal lines have been added to the phase plot at $-110°$ and to the magnitude plot at $20 \log_{10} 1.083 = 0.693$ dB, and a vertical line has been drawn at the frequency 1.42 rad/s.

FIGURE 9.2 *Bode plot for plant and sensor in Example 9.1*

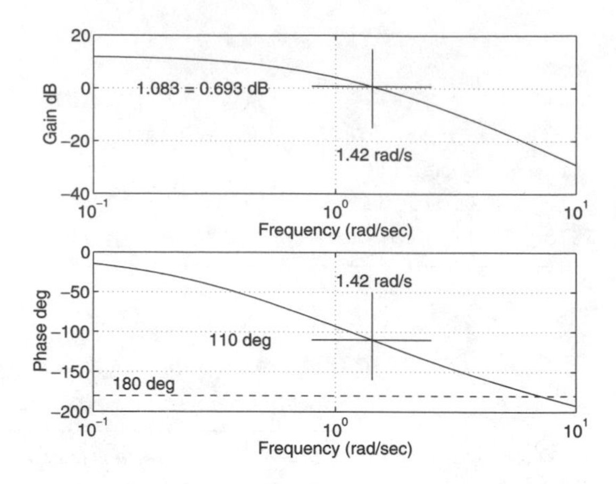

By using the `dcgain` command for the plant-sensor combination and multiplying the result by the value of K_{lag}, we find that the open-loop low-frequency gain is 3.69. This allows us to calculate the value of the zero/pole ratio as $\alpha_{\text{lag}} = 49/3.69 = 13.27$, which will result in an overall low-frequency gain of 49.

b. We use the commands in Script 9.1(b) with $K_{\text{lag}} = 0.924$ and $\alpha_{\text{lag}} = 13.27$, to incorporate the lag compensator, with the value of the lag zero entered by the user via the `input` command. After trying several values in the recommended interval $[\omega_c/10, \omega_c/2]$ where ω_c was found to be 1.42 rad/s in part (a), we learn that a value of $z_{\text{lag}} = -0.25$ results in a phase margin of 60.8°. As this value is very close to our design specification of 60°, we terminate the process.

MATLAB Script

```
% Script 9.1 (b):  select lag zero & calc phase margin
%    Note: variables defined in Script 9.1(a) are in workspace
w_Zlag = input('Enter corner freq for lag zero..... ')
Zlag = -w_Zlag                          % lag zero must be negative
Gc = tf(Klag*[1 -Zlag],[1 -Zlag/Alag])        % lag controller
GcGpH = Gc*GpH        % connect lag in series with plant and sensor
lfg = dcgain(GcGpH)                     % low-freq gain of current design
[km,pm,wkm,wpm] = margin(GcGpH);        % open-loop frequency response
disp([20*log10(km) wkm])           % gain margin in dB & omega_gm
disp([pm wpm])                     % phase margin in degrees & omega_pm
ess = 1/(1 + dcgain(GcGpH))       % verify that steady-state error is OK
```

The open-loop magnitude and phase angle plots for the initial stage (dashed) and for the final design (solid) appear in Figure 9.3. An examination of the solid curves verifies that the phase-margin and low-frequency-gain

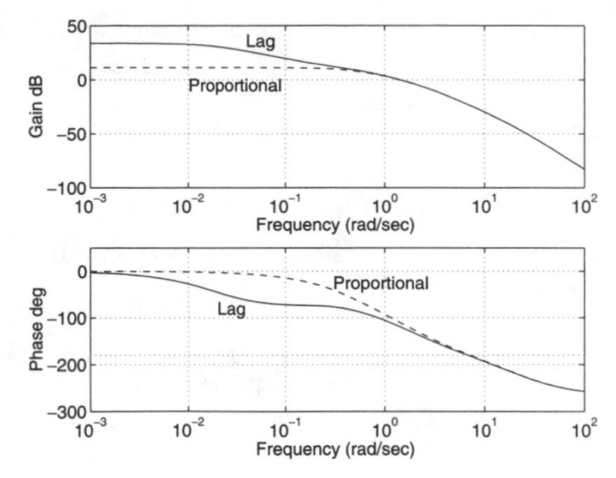

FIGURE 9.3 *Bode plots for initial (proportional) and final (lag) designs in Example 9.1*

specifications have been satisfied. The figure also clearly illustrates the effect of the pole and the zero of the lag compensator on the magnitude and phase angle of the frequency response. At low frequencies the magnitude has been increased by the factor $\alpha_{\text{lag}} = 13.46$, which allows us to meet the steady-state error specification. The phase lag introduced by the compensator has mostly dissipated by $\omega = 1.4$ rad/s, so the phase margin of the final design (solid curve) is 60°, about 10° less than that for the proportional-only controller (dashed curve).

In Figure 9.4 we show the responses to step reference and disturbance inputs, which can be compared with those of the PI design done in Example 8.2 and illustrated in Figure 8.5. By way of comparison, we see that this frequency-response design has resulted in a shorter rise time with lower overshoot for the reference response. However, there is almost 10% undershoot after the initial peak, a transient that decays more slowly than in the PI response, and a steady-state error of 2%. The peak of the disturbance response is lower than that with the PI controller, but the response in Figure 9.4 decays more slowly

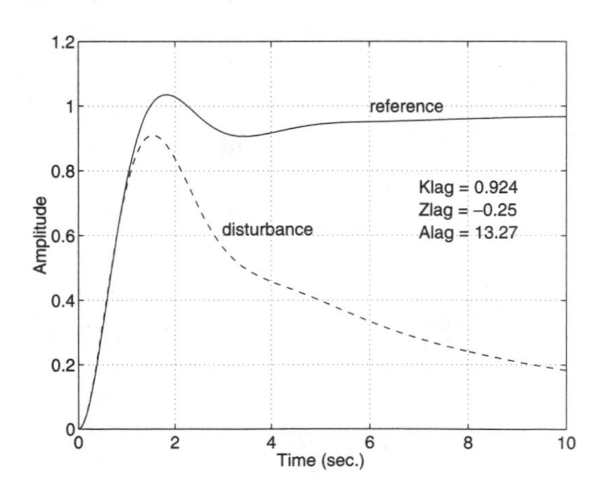

FIGURE 9.4 *Responses to unit steps in the reference and disturbance inputs for the final lag design of Example 9.1*

FREQUENCY-RESPONSE DESIGN

and will not reach zero. These are typical of the tradeoffs that control-system designers must make, and there are no easy ways to resolve them.

METHOD 2—REDUCING MID- AND HIGH-FREQUENCY MAGNITUDE

Kuo (1995) discusses and illustrates another approach to the frequency-response design of lag compensators. It involves starting at the low end of the frequency axis and fixing the gain of the controller at a value that will meet the steady-state error requirement. Then the parameters z_{lag} and α_{lag} are selected to provide the specified phase margin. In contrast with (3.2) and (9.1), the transfer function of the lag compensator is now defined according to

$$G_c(s) = K_{\text{lag}}\left[\frac{-(1/z_{\text{lag}})s + 1}{-(\alpha_{\text{lag}}/z_{\text{lag}})s + 1}\right] \tag{9.3}$$

where $\alpha_{\text{lag}} > 1$ and $z_{\text{lag}} < 0$. When defined this way, $G_c(s)$ will have a low-frequency gain of K_{lag} and a high-frequency gain of $K_{\text{lag}}/\alpha_{\text{lag}}$. Once K_{lag} has been fixed (based on the steady-state error requirement), the frequency-dependent part of the lag will reduce the gain of the combined controller, plant, and sensor, thereby increasing the phase margin. Note that Kuo uses the reciprocal of the zero/pole ratio α_{lag} that we use. Either definition will work, provided that you are consistent.

The steps in the design procedure, as it will be implemented here, are:

1. Determine the controller gain K_{lag} required to satisfy the specification on the low-frequency gain. Then draw a Bode plot of the open-loop system with a proportional-only controller that has the required gain. It could be that the closed-loop system is unstable at this stage, but that will be corrected shortly.
2. The new gain-crossover frequency ω_c is determined by finding the frequency at which the phase-angle curve would result in the desired phase margin (where an extra 5° to 10° has been added) if the magnitude curve were to pass through 0 dB (or unity gain) at that frequency. We use the magnitude of $K_{\text{lag}}G_p(j\omega_c)H(j\omega_c)$ to determine the value of α_{lag} required to reduce the magnitude of the compensated system at this frequency to 0 dB by requiring that

$$\alpha_{\text{lag}} = |K_{\text{lag}}G_p(j\omega_c)H(j\omega_c)| \tag{9.4}$$

3. We select the corner frequency of the lag zero to be one decade below ω_c so $z_{\text{lag}} = -\omega_c/10$. The compensator pole will be at $s = z_{\text{lag}}/\alpha_{\text{lag}}$. At this point the design is complete, unless we wish

to iterate on one or more of the design parameters after examining the responses of the system to reference and disturbance step inputs.

Next, we will apply this design algorithm to the same plant and sensor combination used in the previous example, with the same specifications on steady-state error and phase margin.

EXAMPLE 9.2 *Lag Controller Design, Method 2*

Using the frequency-response design algorithm described above, find a lag compensator that will result in a phase margin of approximately 60° and an open-loop low-frequency gain of 49 for the plant and sensor whose transfer functions are given in (9.2).

Solution

a. We commence by building the series interconnection of the gain K_{lag}, the open-loop plant $G_p(s)$, and the sensor $H(s)$. The controller gain is calculated as $K_{\text{lag}} = 49/4 = 12.25$, which will provide the low-frequency gain required and will result in a steady-state error of 2% for a step-reference input. The Bode plot for this system is shown in Figure 9.5. It is clear that there is a phase margin of only several degrees, because at the frequency for which the magnitude is 0 dB ($\omega_{\text{pm}} = 6.7$ rad/s) the phase angle is close to $-180°$. The actual phase margin as given by the `margin` command is 2.6°. Hence the closed-loop system is almost marginally stable unless the magnitude curve can be lowered for higher frequencies so as to significantly increase the phase margin.

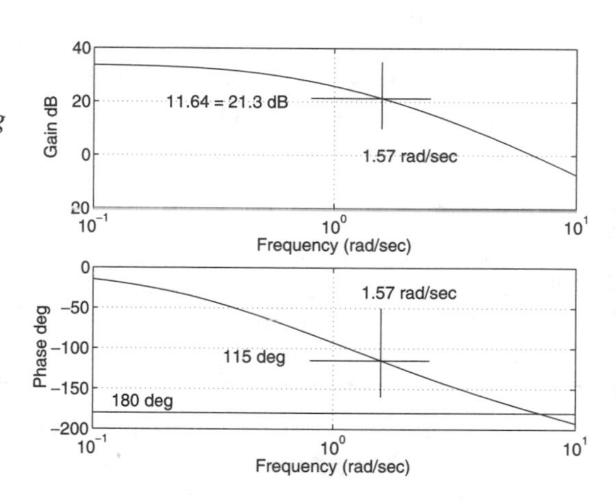

FIGURE 9.5 *Bode plot for Example 9.2 with proportional control having* $K_{\text{lag}} = 12.25$

b. We will look for a phase margin of 65° (the 60° that we need plus an extra 5°). For this to be the case, the phase at the new gain-crossover frequency must be $-180 + 65 = -115°$. We can see from the phase curve in Figure 9.5 that the new gain-crossover frequency will be close to $\omega = 1.5$ rad/s. Rather than having to estimate ω_c from the Bode plot, we

FREQUENCY-RESPONSE DESIGN

use the MATLAB command interp1 to interpolate the function $\omega(\phi)$ for the frequency at which the phase angle is $-115°$. The instruction to accomplish this is wc = interp1(ph,w,-115), which results in $\omega = 1.57$ rad/s. The magnitude of the open-loop frequency response at this frequency is 11.64, which is equivalent to 21.3 dB.

c. Next, the bode command in Script 9.2(c) is used at the single frequency ω_c to evaluate the right-hand side of (9.4), which gives the value of the zero-pole ratio α_{lag}. The corner frequency of the lag zero is taken as $\omega_c/10$. At this point all three of the controller parameters have been established, and we can build the TF model of the lag controller and use the * operator to connect it with the models of the plant and the sensor. Finally, the margin command can be used to determine the actual phase margin attained. If the value of ϕ_m is not quite what was desired, the commands in the script can be rerun with a different value entered for ω_c, where a lower value of the crossover frequency should result in a larger value of α_{lag} and a greater phase margin.

For the example under consideration, we settle on $\omega_c = 1.57$ rad/s, which results in a phase margin of 59.6°, which is just about what we are looking for. Using this value, we get $\alpha_{\text{lag}} = 11.64$ and $z_{\text{lag}} = -0.157$.

MATLAB Script

```
% Script 9.2(c):  Calculate alpha and lag zero for Kuo's method
%  Use GpH from Script 9.1(a)
%  Klag has been determined in part (a) of this example
%  gain-crossover freq wc was found in part (b)
[mag_KGpH,ph_KGpH] = bode(Klag*GpH,wc)          % calc only at wc
Alag = mag_KGpH                                 % lag alpha per (9.4)
Zlag = -wc/10                                   % lag zero
%-- using Kuo's form for lag (unity low-freq gain) rather than (3.2)
Gc = tf(Klag*[-1/Zlag 1],[-Alag/Zlag 1]);
GcGpH = Gc*GpH;                           % open-loop compensated system
lfg = dcgain(GcGpH)                       % low-freq gain of current design
[km,pm,wkm,wpm] = margin(GcGpH);          % open-loop frequency response
```

This completes the design, with the proviso that one or more of the controller's three parameters may be adjusted following inspection of the step responses of the closed-loop system, as depicted in Figure 9.6. The open-loop magnitude and phase-angle plots for the final design are the solid curves in Figure 9.7. It is apparent that the lag has reduced the magnitude for $\omega > 0.02$ rad/s, thereby allowing the phase margin specification to be satisfied. In the lower plot we see that the phase lag due to the controller has almost dissipated by the gain-crossover frequency ω_c. A comparison of the

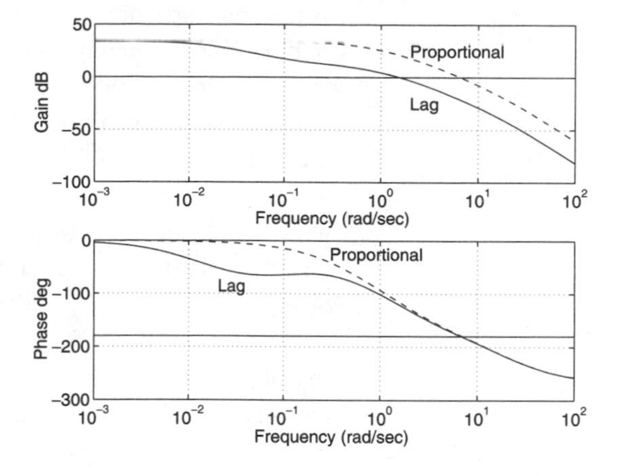

FIGURE 9.6 *Bode plots for initial (proportional) and final lag designs of Example 9.2*

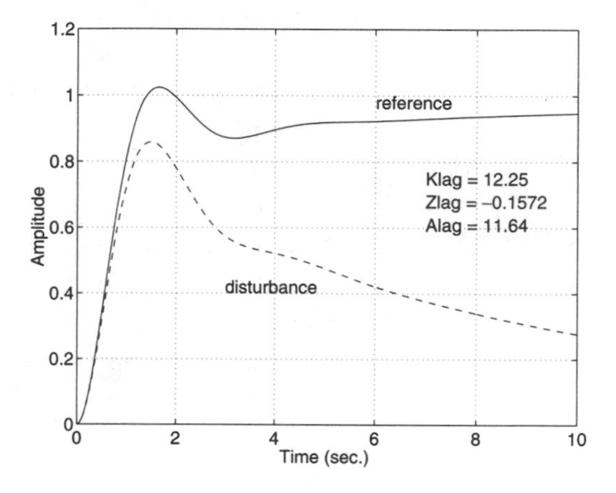

FIGURE 9.7 *Responses to unit steps in the reference and disturbance inputs for the final lag design of Example 9.2*

magnitude curve with that in Figure 9.3, which was obtained with Method 1, shows no discernible difference. However, there are some small differences, say up to 10° in the two phase curves. For the design done with Method 2, the phase begins to decrease at a lower frequency and actually increases slightly before eventually falling to $-270°$. This earlier decrease is because its zero ($s = -0.157$) has a smaller magnitude than the zero of the design done with Method 1 ($s = -0.250$).

LEAD CONTROLLER DESIGN

Like its counterpart, the proportional-plus-derivative (PD) controller, a lead controller is generally used for reducing the rise time or, equivalently, increasing the bandwidth of the closed-loop frequency response for the feedback system shown in Figure 9.1. As we did with the lag controller, we assume that specifications are to be met for steady-state error, which will establish the minimum required low-frequency gain, and for phase margin, which should ensure adequate damping. Keep in mind that, although we will be satisfying a specification on the steady-state error due to a reference-step input, only the gain of the lead is being used to meet that requirement. The dynamic aspects of the lead, as established by its zero and pole (or zero and alpha) are being used to provide the required phase margin, subject to the constraint that the lead gain has been established based on steady-state error considerations.

The method described by Kuo (1995), which is essentially the same as that presented by Franklin, Powell, and Emami-Naeini (1994), begins by fixing the controller gain K_{lead} at the value which will provide the required low-frequency gain. Presumably this gain results in a phase margin that is too low to meet the specification and yields a step response that is too lightly damped. The system could even be unstable at this point in the design process.

To rectify this situation, a lead compensator is added that has the transfer function

$$G_c(s) = K_{\text{lead}}\alpha_{\text{lead}}\left(\frac{s - z_{\text{lead}}}{s - \alpha_{\text{lead}} z_{\text{lead}}}\right) \tag{9.5}$$

This transfer function differs from that given by (3.3) in that the gain has been defined here to make the low-frequency gain of the lead be just K_{lead}, rather than $K_{\text{lead}}/\alpha_{\text{lead}}$ as it is for (3.3). The lead compensator, or controller, provides phase lead and increases the mid- and high-frequency magnitudes. Whereas the former is beneficial to the phase margin, the latter is not. Thus, the designer must select parameter values (z_{lead} and α_{lead}) to balance the positive and negative effects.

After the gain K_{lead} has been added, we use MATLAB to compute the modified phase margin, expecting that it will not satisfy the specification. Then we decide on a phase margin target value, realizing that we will not end up with anything close to that value because the magnitude-versus-frequency curve generally has a negative slope, and the gain-crossover frequency ω_c is going to increase substantially due to the lead. The hoped-for increase in the phase margin, denoted as $\Delta\phi_m$, is the difference between the user-specified target value and the actual phase margin with just K_{lead} included. It is related to the lead's pole-zero ratio according to

$$\alpha_{\text{lead}} = \frac{1 + \sin\Delta\phi_m}{1 - \sin\Delta\phi_m} \tag{9.6}$$

which we can use to solve for α_{lead}. Also, we know that the maximum phase lead will occur at the lead's center frequency

$$\omega_{\text{ctr}} = z_{\text{lead}}\sqrt{\alpha_{\text{lead}}} \tag{9.7}$$

and at this frequency the magnitude of the lead's frequency response is $\sqrt{\alpha_{\text{lead}}}$.

At this point, we use the MATLAB command `interp1` with the open-loop frequency response data obtained with only K_{lead} as the controller to solve for the center frequency of the lead ω_{ctr} to satisfy the relationship

$$|K_{\text{lead}} G_p H(j\omega_{\text{ctr}})| = 1/\sqrt{\alpha_{\text{lead}}} \tag{9.8}$$

If we do this, ω_{ctr} will become the gain-crossover frequency when the lead is installed, because the magnitude of the series combination of the lead and the plant (including any sensor) will be unity (or 0 dB) at this frequency. Then we can use (9.7) to solve for the lead zero.

When we compute the phase margin of the system with the lead compensator in place, we generally find that we achieve only a fraction of the increase in phase margin that we had asked for. This partial success is because the gain-crossover frequency ω_c is being moved to a higher frequency, and the phase of the plant's frequency response is decreasing. Thus, in practice, we ask for a somewhat larger increase in phase margin than we will really need and expect to make several attempts before coming up with a satisfactory choice. We will illustrate these points in the example that follows.

EXAMPLE 9.3 ***Lead Controller Design***

Use the frequency-response approach described above to design a lead compensator for the plant and sensor whose transfer functions are given by (9.2) such that the closed-loop system will have a steady-state error of 4% to a unit-step reference input and its phase margin will be approximately 60°.

Solution

The solution of this example is divided into the four steps described below, and the MATLAB commands that will implement these steps are given in Script 9.3 at the end.

a. Because the plant and sensor constitute a type-0 system, the steady-state error specification of 4% = 1/25 dictates a position constant, as defined by (7.3), of $K_p = 25 - 1 = 24$. The low-frequency gain of the plant and sensor is 4, so the lead must supply a gain of $24/4 = 6$ at low frequencies, which requires that $K_{\text{lead}} = 6$.

b. Next, we compute the phase margin with the proportional-only controller $G_c(s) = K_{\text{lead}}$ and find it to be 16.6°, which is well below the specification of 60°. To compensate for the decreasing magnitude and phase of the plant in the gain-crossover region, we ask for a total phase margin of 85°, which means an increase of $\Delta\phi_m = 85 - 16.6 = 68.4°$. Then, (9.6) can be used to solve for $\alpha_{\text{lead}} = 27.52$.

c. We want the center frequency of the lead to be the new gain crossover frequency ω_c so we look for that frequency for which the magnitude response of the open-loop system with the gain $K_{\text{lead}} = 6$ included is $1/\sqrt{\alpha_{\text{lead}}} = 0.1906$. By doing an interpolation with the vectors `w` and `mag` produced with the bode command, we find this frequency to be ω_{ctr} = 10.47 rad/s. Based on these values for the center frequency and for α_{lead}, the zero and pole of the lead compensator are $z_{lead} = -\omega_c/\sqrt{\alpha_{lead}} = -1.995$ and $p_{lead} = z_{lead}\alpha_{lead} = -54.90$. We compute the actual phase margin as 54.4°, which is 5.6° less than what we are looking for.

Because the value for α_{lead} is on the high end of our acceptable range, we will not attempt to improve the phase margin. In fact, using values larger than 85° for the phase-margin target will result in much larger values for α_{lead} but not much of an increase in ϕ_m, and maybe a decrease in the phase margin. At this point we are getting all we can from the single lead, and to obtain a larger phase margin we need to (i) reduce the low-frequency gain, thereby incurring a larger steady-state error, or (ii) increase the complexity of the controller by adding either another lead section or a lag. The last option is the subject of the next section, so we will keep the controller as it stands. The Bode plots in Figure 9.8 show how the lead has increased the magnitude at the higher frequences and has increased the phase angle over the mid to high frequences. The maximum increase in phase occurs at $\omega_c = 10.47$ rad/s, which is the lead's center frequency.

d. We complete the design process by plotting the responses of the closed-loop system to unit-step reference and disturbance inputs, resulting in Figure 9.9. When these responses are compared with those in Figures 9.3 and 9.6, which were obtained with a lag controller and have steady-state errors of only 2% to the step-reference input, we see that the lead controller has resulted in an appreciably faster response but twice the steady-state error. In general, we would use a lead to get a faster response and a lag to get lower steady-state errors. In the next two sections we will illustrate two different approaches to designing lead-lag controllers that combine these features and obtain corresponding improvements in both the reference and disturbance step responses.

MATLAB Script

```
% Script 9.3 -- lead controller design
% use GpH & tau_sen from Script 9.1(a)
%----- Part (a) determine lead gain -----
Kp_reqd = (1/0.04) - 1              % required gain to give 4% ess
plant_lfg = dcgain(GpH)
Klead = Kp_reqd/plant_lfg
%----- Part (b) determine lead alpha -----
w = logspace(-1,2,100);
[mag,ph] = bode(Klead*GpH,w);
mag = reshape(mag,100,1);
```

Continues

```
ph = reshape(ph,100,1);
pm_target = 85                      % phase-margin target in degrees
del_pm = pm_target - pm             % max phase angle in degrees
Alead = (1+sin(del_pm*pi/180))/(1-sin(del_pm*pi/180)) % Eq (9.6)
%----- Part (c) determine lead zero -----
wc = interp1(mag,w,1/sqrt(Alead))       % interpolate (9.8) to find wc
Zlead = -wc/sqrt(Alead)                               % using (9.7)
Gc = tf(Klead*Alead*[1 -Zlead],[1 -Alead*Zlead])  % use (9.5) for lead TF
GcGpH = Gc*GpH;        % OL transfer function of lead, plant, & sensor
[km,pm,wkm,wpm] = margin(GcGpH)         % find new phase margin
%----- Part (d) CL system for reference input with sensor in fdbk path ----
T_ref = feedback(Gp*Gc,H,-1)
```

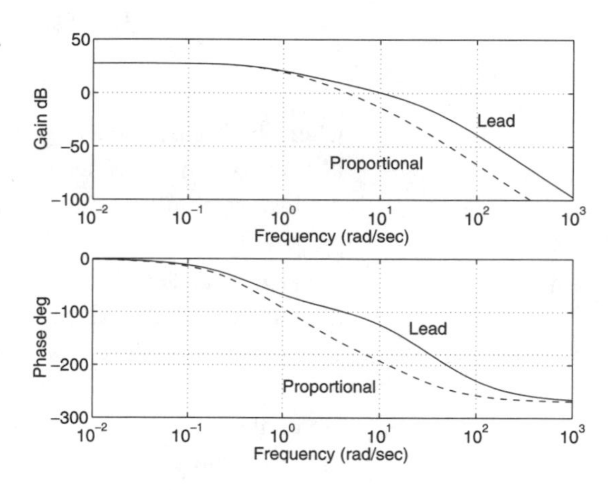

FIGURE 9.8 *Bode plots for initial (proportional) and final lead designs of Example 9.3*

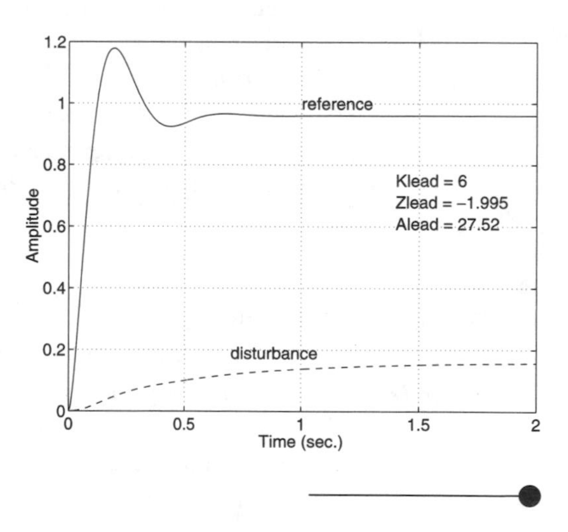

FIGURE 9.9 *Responses to unit steps in the reference and disturbance inputs for the final lead design of Example 9.3*

LEAD-LAG CONTROLLER DESIGN USING FREQUENCY RESPONSE

Typically, we will need to do a lead-lag design when we require a small steady-state error, a short rise time, and a small overshoot. It will generally not be possible to meet such specifications with either a lead or a lag alone, but we can often achieve success if both features are included in the controller. The transfer function of the lead-lag controller was given in (3.4), where the same value of α was used for both the lead and lag portions. Here we wish to remove that constraint, so we write the controller transfer function as

$$G_c(s) = K_{\text{ldlg}}\left(\frac{s - z_{\text{lag}}}{s - z_{\text{lag}}/\alpha_{\text{lag}}}\right)\left(\frac{s - z_{\text{lead}}}{s - \alpha_{\text{lead}} z_{\text{lead}}}\right) \tag{9.9}$$

where both α_{lead} and α_{lag} are greater than unity.

We will have to decide on the controller gain (K_{ldlg}), the zero and α for the lead section, and the zero and α for the lag section. We would expect that $z_{\text{lead}} < z_{\text{lag}} < 0$.

To avoid dealing with too many unknown parameters at a time, we will first do a lead design according to the steps outlined and illustrated in the previous section. Then we will add a lag compensator to the system that we have at the conclusion of the lead design step. For this, we will use method 2, in which the lag with unity low-frequency gain is used to attenuate the magnitude values in the mid- and high-frequency range. As before, we will try to meet specifications on the low-frequency gain K_p if the plant is type-0, to satisfy a requirement on the steady-state error to a step input. For a type-1 plant, the corresponding specification will be on the velocity constant K_v which governs the steady-state error to a ramp input. We will also try to attain a minimum value for the phase margin to ensure a satisfactory overshoot for the reference step response. To ensure a short rise time for the reference step response, we will also try to meet a specification on the bandwidth of the closed-loop system.

EXAMPLE 9.4
Lead-Lag Controller Design

For the plant and sensor transfer functions given in (9.2) that have been used in earlier design examples, find a lead-lag controller that will result in a low-frequency gain of 400 and a phase margin of at least 50°. The values of α for both sections should not exceed 40. After the design has been completed, determine the transfer function of the closed-loop system and find its bandwidth.

Solution The solution of this example is divided into the four steps described below, and the MATLAB commands that will implement these steps are given in Script 9.4 that follows.

a. Because $G_p(0) = 4$ and $H(0) = 1$, the controller gain required to meet the low-frequency-gain specification is $K_{\text{ldlg}} = 400/4 = 100$. When we compute the frequency response of the plant and sensor, with this gain we find that the phase margin is $\phi_m = -32.7°$, which means that the closed-loop system without the lead and lag sections is unstable.

b. Next, we insert a lead section that has unity low-frequency gain and follow the same steps that we employed for the lead design in Example 9.3. We use a value of 35° for the target phase margin and find that $\Delta\phi_m = 67.7°$, which requires that $\alpha_{\text{lead}} = 25.66$. With the `interp1` command we use (9.8) to perform an interpolation on the frequency-response data with only K_{ldlg} for the controller to find that the center frequency for the lead should be 32.59 rad/s. From (9.7) it follows that the lead zero should be at $s = -6.43$ and means that the lead pole is at $s = -6.43 \times 25.66 = -165$, which is well to the left in the s-plane. With the lead installed, the `margin` command indicates that the phase margin at this stage of the design has been improved from $-32.7°$ (unstable) to $13.7°$ (stable, but lightly damped).

c. To meet our specified phase margin of 50°, we add a lag section by following the steps used in Example 9.2. Because the controller gain has already been fixed, we start at step (b) of the example and request a phase margin that is a bit greater than our goal of 50°. Then we use the `interp1` command to solve for the gain-crossover frequency ω_c, from which we can calculate the zero-pole ratio and the zero of the lag. Finally we use the `margin` command to determine what phase margin we attained. Assuming that we do not get just what we want the first time, we repeat the process with a different phase-margin request and recompute the lag parameters α_{lag} and z_{lag}.

After several iterations we find that, if we make the new gain-crossover frequency be that frequency for which the phase angle is $\phi = -124°$, we will satisfy the phase-margin specification with $\phi_m = 50.4°$. The corresponding lag parameters are $\alpha_{\text{lag}} = 33.2$ and $z_{\text{lag}} = -0.338$. Figure 9.10 shows the open-loop frequency responses after the initial stage (dashed) and for the final design (solid). The upper plot shows how the controller has reduced the magnitude over a wide range of frequencies. The lower plot shows the effects of the lag in reducing the phase angle for lower frequencies and the effect of the lead in increasing the phase for higher frequencies.

d. The last block of commands in the script computes the bandwidth of the closed-loop system. To accomplish this we: (i) use the * and `feedback` commands to build the TF model of the closed-loop system, (ii) select a one-decade wide range of frequencies that will include the point at which the magnitude becomes 3 decibels below the low-frequency value and evaluate the closed-loop frequency response, (iii) use the `dcgain` command to find the low-frequency gain (-0.022 dB), and (iv) use the RPI function `bwcalc` to compute the final value of 5.28 rad/s. One might be tempted to use the interpolation command `interp1` instead, but the magnitude function `mag_CL_db` is not monotonic, which prevents the use of `interp1` unless the range of frequencies is suitably restricted.

The reference and disturbance step responses for the final design are shown in Figure 9.11. Comparing the reference step response with that of Figure 8.8, which was obtained with a PID controller, we see that in this case there is more overshoot and a longer settling time. This design will have a steady-state error of $1/401 = 0.25\%$, whereas the PID design from Chapter 8 will have zero steady-state error. However, no effort was made to optimize this design by making different choices in the design process.

MATLAB Script

```
% Script 9.4 lead-lag controller design
% use GpH & tau_sen from Script 9.1(a)
%----- Part (a) determine gain & check phase margin -------
Kp_reqd = 400         % will give ess = 1/401
plant_lfg = dcgain(GpH)
Kldlg = 400/plant_lfg
[km,pm,wkm,wpm] = margin(Kldlg*GpH)                  % calc pm with gain only
%----- Part (b) insert lead, determine its alpha & zero ------
w = logspace(-1,2,100)';                % 100 freq values between 0.1 & 100 rps
[mag,ph] = bode(Kldlg*GpH,w);
mag = reshape(mag,100,1);                   % convert results to 100 × 1 vectors
ph = reshape(ph,100,1);
pm_target = input('enter phase-margin target in deg....')
del_pm = pm_target - pm                     % max phase angle in deg
Alead = (1+sin(del_pm*pi/180))/(1-sin(del_pm*pi/180))          % Eq (9.6)
wc = interp1(mag,w,1/sqrt(Alead))  % interpolate (9.8) to find wc
Zlead = -wc/sqrt(Alead)                % using (9.7)
Gld = tf(Kldlg*Alead*[1 -Zlead],[1 -Alead*Zlead]); % use (9.5) for lead TF
% OL transfer function of lead, plant, & sensor
[km,pm,wkm,wpm] = margin(GpH*Gld);  % new pm value
%------ Part (c) insert lag & determine its alpha & zero -----------
[mag_ld,ph_ld]  = bode(GpH*Gld,w);
mag_ld = reshape(mag_ld,100,1); ph_ld = reshape(ph_ld,100,1);
ph_at_wc = input('enter desired phase at new wc in deg....')
wc = interp1(ph_ld,w,ph_at_wc)            % gain crossover freq
mag_ld_wc = bode(GpH*Gld,wc)              % use only one freq
```

Continues

```
Alag = mag_ld_wc                               % lag alpha
Zlag = -wc/10                                  % lag zero
Glg = tf([-1/Zlag  1],[-Alag/Zlag  1]) % use Kuo's form for unity DC gain lag
Gc = Gld*Glg              % combine lead & lag for final lead-lag controller
GcGpH = Gc*GpH               % put controller in series with plant + sensor
[km,pm,wkm,wpm] = margin(GcGpH);       % want pm = 50
%--- Part (d) build CL system for reference input with sensor in fdbk path
T_ref = feedback(Gp*Gc,H,-1)
%---- bandwidth of CL system using narrow freq interval
ww = logspace(0,1,100)';                          % 1-decade freq interval
[mag_CL,ph_CL] = bode(T_ref,ww);                 % magnitude as ratio
mag_CL = reshape(mag_CL,100,1);
mag_CL_db = 20*log10(mag_CL);                    % convert mag to dB
lfg = dcgain(T_ref)                              % low-freq gain as ratio
lfg_db = 20*log10(lfg)                           % low-freq gain in dB
bw_CL = bwcalc(mag_CL_db,ww,lfg_db)              % calc CL bandwidth in rad/s
```

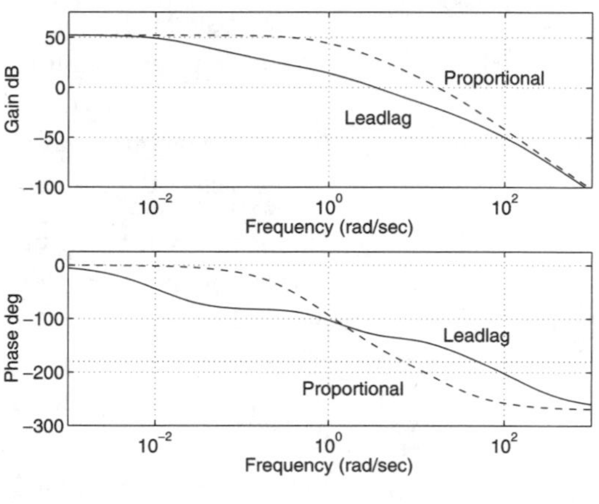

FIGURE 9.10 *Bode plots for initial (proportional) and final lead-lag designs of Example 9.4*

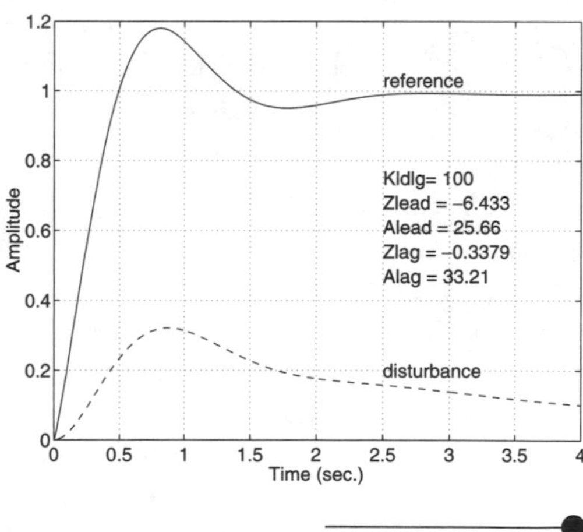

FIGURE 9.11 *Responses to unit steps in the reference and disturbance inputs for the final lead-lag design of Example 9.4*

FREQUENCY-RESPONSE DESIGN

LEAD-LAG CONTROLLER DESIGN BASED ON PID DESIGN

In Section 8.3 we defined the PID control law by writing a time-domain equation and showing that its transfer function (8.2) was

$$G_c(s) = K_P\left(\frac{K_D s^2 + s + K_I}{s}\right) \tag{9.10}$$

This transfer function has a pole at $s = 0$ and, provided that $K_D K_I < 0.25$, a pair of real zeros. We will create a lead-lag controller whose transfer function (9.9) has the same two real zeros as the PID transfer function, two real poles, a single pole-zero ratio $\alpha_{\text{lead}} = \alpha_{\text{lag}} = \alpha_{\text{ldlg}}$, and a gain. The design will be implemented by doing the following:

1. Make the lag zero equal to the zero of the PID transfer function (9.10) that is closest to the origin of the s-plane, and make the lag pole be $1/\alpha_{\text{ldlg}}$ times the lag zero, where we select the value of α_{ldlg} somewhat arbitrarily to a number like 10, 15, or 20;
2. make the lead zero be the zero of (9.10) that is farthest from the origin, and make the lead pole be α_{ldlg} times the lead zero; and
3. select the gain according to a step-response overshoot specification.

Doing this will result in the controller transfer function $G_c(s)$ given by (9.9) where, for simplicity, both the lead and lag portions use the same value of α, denoted by α_{ldlg}.

Figure 9.12 shows the pole-zero representations of the two controllers in the s-plane corresponding to a hypothetical PID controller. Assuming that we have values for K_I and K_D from a PID design, we can solve for the zeros of the lead-lag transfer function and compute the lead and lag poles for a specified value of α_{ldlg}. To determine the gain of the lead-lag controller K_{ldlg}, we will use MATLAB to perform a gain sweep on the step response of the closed-loop system to a reference input and select a value that gives satisfactory overshoot, steady-state error, and settling-time properties.

EXAMPLE 9.5
Lead-Lag Control Based on PID Design

Design a lead-lag controller for the process and sensor whose transfer functions $G_p(s)$ and $H(s)$ are given in (9.2). The parameters of the lead-lag controller are to be determined according to the procedure described above, using the results of the PID controller designed in Example 8.3, for which $K_P = 7, K_I = 0.5$, and $K_D = 0.4$. The specific steps to be taken are:

a. Calculate the lead-lag controller zeros z_{lag} and z_{lead} to be the same as the zeros corresponding to the PID controller parameters given above. Then use $\alpha_{\text{ldlg}} = 20$ to determine the values of the two poles.

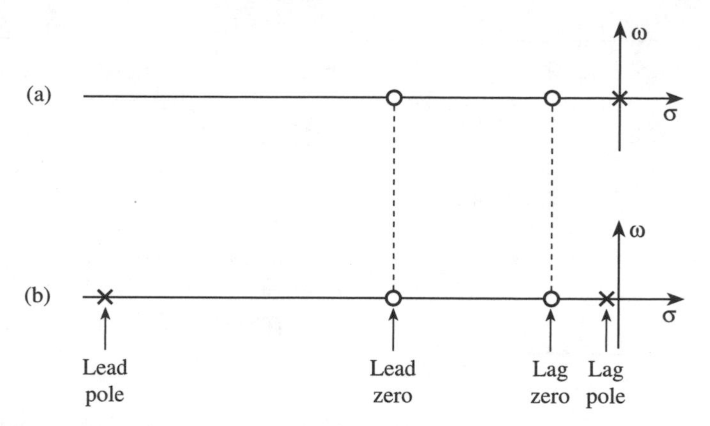

FIGURE 9.12 *Controller pole-zero diagrams (a) PID (b) Lead-lag*

b. Draw the root locus for variations in K_{ldlg} and use the `rlocfind` command to determine K^*_{ldlg}, the maximum gain for stability. Alternatively, use the `margin` command to obtain the gain margin and use it to calculate K^*_{ldlg}.

c. Use MATLAB to compute and plot the response of the closed-loop system to a unit step in the reference input, where the controller gain K_{ldlg} is swept over a suitable range of values whose maximum is less than K^*_{ldlg}. On the basis of the results, select a gain value that provides a good balance between overshoot, steady-state error, and settling time.

d. Using the value for K_{ldlg} selected in the previous step, plot the responses to step reference and disturbance inputs, and compare your results with those in Figure 8.5(b). Determine the steady-state errors to both inputs and explain why they are not zero. Recall that with the PID controller both the steady-state errors were zero.

Solution

a. We can use Script 9.5, where $K_I = 0.5$ and $K_D = 0.4$, to show that when the zeros of the lead-lag controller are made to agree with those of the PID controller we get $z_{\text{lag}} = -0.691$ and $z_{\text{lead}} = -1.809$. If we take $\alpha_{\text{ldlg}} = 20$ and leave the gain in symbolic form, we can write the transfer function of the lead-lag controller as

$$G_{\text{ldlg}}(s) = K_{\text{ldlg}}\left(\frac{s + 0.691}{s + 0.691/20}\right)\left(\frac{s + 1.809}{s + 1.809 \times 20}\right)$$
$$= K_{\text{ldlg}}\left(\frac{s + 0.691}{s + 0.0345}\right)\left(\frac{s + 1.809}{s + 36.18}\right)$$

MATLAB Script

```
% Script 9.5 : Lead-lag zeros that match PID zeros
KI = 0.5, KD = 0.4                  % PID controller parameters
Alpha = 20                          % ratio of lead-lag zero to pole
Zpid = roots([KD  1  KI])           % both should real & negative
Zlead = min(Zpid)                   % farthest to left becomes lead zero
Zlag = max(Zpid)                    % closest to origin becomes lag zero
numGldlg = Kldlg*conv([1 -Zlag],[1 -Zlead])  % num of GC(s)
denGldlg =  conv([1 -Zlag/Alpha],[1 -Alpha*Zlead]) % denominator
Gldlg = tf(numGldlg,denGldlg)          % TF of lead-lag Gc(s)
```

b. To obtain the root locus, we modify Script 8.1(a) to implement the lead-lag controller transfer function given above but with K_{ldlg} omitted or set to unity. Doing this, we find that the maximum value of the controller gain for which the closed-loop system will be stable is $K_{\mathrm{ldlg}}^{*} = 519.8$.

The corresponding gain margin is 54.15 dB, which means that marginal stability will result for $K_{\mathrm{ldlg}} = 54.15$ dB. Converting this value to a magnitude ratio gives $K_{\mathrm{ldlg}}^{*} = 10^{(54.15/20)} = 10^{2.7075} = 509.9$, which is reasonably close to the value of 519.8 that was obtained from the root-locus plot.

c. The results of the gain sweep are shown in Fig. 9.13. It is apparent that for $K_{\mathrm{ldlg}} = 70$ the step response will result in approximately 10% overshoot and only a very small steady-state error.

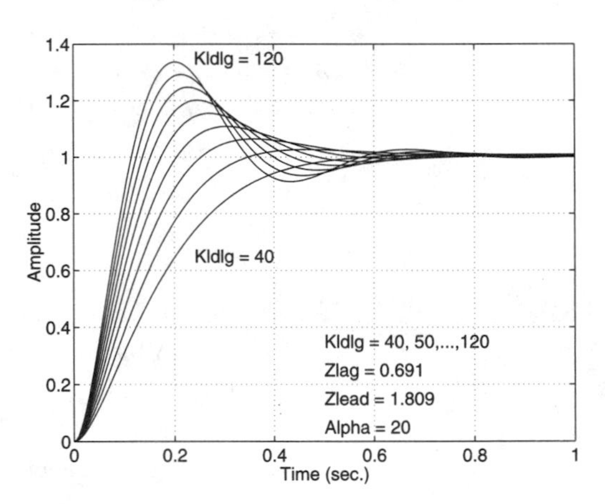

FIGURE 9.13 *Responses to unit steps in the reference input with K_{ldlg} swept from 40 to 120 in increments of 10*

d. Using the final set of parameters for the lead-lag controller, namely $K_{\mathrm{ldlg}} = 70$, $z_{\mathrm{lag}} = -0.691$, $z_{\mathrm{lead}} = -1.809$, and $\alpha_{\mathrm{ldlg}} = 20$, we obtain the step responses for the reference and disturbance inputs shown in Figure 9.14.

FIGURE 9.14 *Responses to unit steps in the reference and disturbance inputs with $K_{\text{ldlg}} = 70$*

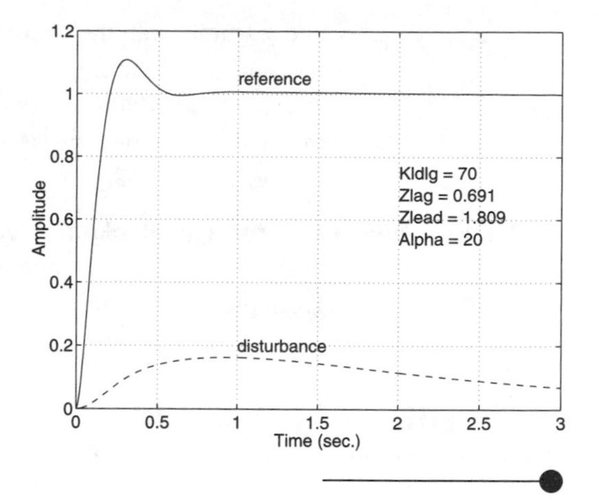

Comment: It is interesting to compare these responses with those in Figure 8.8 that were obtained with the PID controller and with those in Figure 9.12 from the lead-lag controller designed with the frequency-response methods. The reference-input step response with the lead-lag controller based on the PID design is not quite as fast as that of the PID design but is much faster than the other lead-lag design, and with less overshoot. For the disturbance input, the three controllers rank in the same order, with the differences being in the maximum value of the error. However, in comparing these responses, one should keep in mind that the amounts of optimization employed in arriving at the final parameter values of the three designs were not equal. The PID design of Example 8.3 had extensive parameter optimization done, whereas the lead-lag design of this example had some and the other lead-lag design had none. Also, it should be remembered that the PID design is not physically realizable unless a pole is added to the transfer function of the derivative term.

Comment: Although the PID controller is an idealization that cannot be implemented exactly, it is helpful to be able to consider it as a means of (i) establishing the zero locations of the lead-lag design and (ii) providing response measures that can be used as a guide for the level of performance that can be achieved.

EXPLORATION

E9.1 Lag compensator for type-0 plant, Method 1. Consider the 4-th order plant having the transfer function

$$G_p(s) = \frac{160(s + 10)}{(s + 2)(s + 4)(s + 5)(s + 20)}$$

and assume that the feedback transfer function is unity (the dynamics of the sensor can be neglected). Using the frequency-response design approach referred to as Method 1 and illustrated in Example 9.1, design a lag controller that will result in a phase margin of at least 60° and a steady-state error of no more than 2%.

E9.2 Lag compensator for type-0 plant, Method 2. Repeat the previous problem using the frequency-response design approach referred to as Method 2, as illustrated in Example 9.2. Try to satisfy the same specifications.

COMPREHENSIVE PROBLEMS

CP9.1 Electric power generation system. Use the design approaches discussed in this chapter to obtain lag controllers for the voltage control system shown in Figure A.2, in Appendix A. As in Comprehensive Problem CP8.1, assume that some supplementary damping has been added to the lightly damped electromechanical mode, and use the file `epow2.m` to obtain the plant model to start the design.

a. Use Method 1 to design a lag controller. The design should satisfy the specifications that the steady-state error to a unit-step input V_{ref} is no more than 0.002, and the phase margin is at least 70°. Plot the step reference response of the closed-loop system. From the step response, verify the steady-state error, and find the rise time, maximum overshoot, and 2% settling time.
b. Repeat Part (a), using Method 2 to design the lag controller.
c. Adapt the guidelines specified in Section 9.4 to map the PI controller obtained from Comprehensive Problem CP8.1 into a lag controller. Take the lag-controller zero to be that of the PI controller, and set K_{lag} to be K_P. Vary α_{lag} and find its effect on the closed-loop system reference step response in terms of the steady-state error, rise time, maximum overshoot, and 2% settling time. Find the smallest value of α_{lag} that will satisfy the specifications in part (a).

CP9.2 Satellite with TF model. Obtain the model of the open-loop satellite by running the MATLAB file `sat.m` and selecting the transfer function from the motor torque to the pointing angle. Use the frequency-response design method illustrated in Example 9.3 to design a lead controller that will result in a phase margin of at least 50°.

For this system the controller gain must be negative and its magnitude should be kept small to avoid large excursions in the speed of the reaction wheel. The value $K_{\text{lead}} = -0.001$ should prove satisfactory, so start with it. You may wish to consider other values after you have an initial design.

Recall that this form of the satellite model is subject to the restriction that the initial wheel speed is zero. Because of this restriction, a lead controller will result in zero steady-state error to a step change in the desired pointing angle.

SUMMARY

In this chapter we have shown how MATLAB can be used to carry out the rather involved calculations that are required to design lead-lag controllers using frequency-response methods. Two different but related methods were used to design a lag compensator, and a single method has been demonstrated for the lead. Then lead-lag compensators were developed by combining one of the lag methods with that for the lead. Finally, a totally different approach to designing a lead-lag compensator that is derived from a PID design was illustrated. One should keep in mind the fact that although MATLAB can be of great assistance in performing the extensive calculations required for these designs, its use does not diminish the need for human judgment and a considerable amount of trial and error.

MATLAB FUNCTIONS USED

Function	*Purpose and Use*	*Toolbox*
*	Given two LTI objects, the * operator forms their series connection.	Control System
bode	Given a model in TF or SS form, **bode** returns the magnitude and phase of the frequency response. When the output variables are omitted, it generates the Bode plot directly.	Control System
bwcalc	Given the magnitude of the frequency response of a system and its low frequency gain, **bwcalc** computes the bandwidth.	RPI function
conv	Given two row vectors containing the coefficients of two polynomials, **conv** returns a row vector containing the coefficients of the product of the two polynomials.	MATLAB
dcgain	Given a model in TF form, **dcgain** returns the steady-state gain of the system.	Control System

feedback	Given the models of two systems in TF form, **feedback** returns the model of the closed-loop system, where negative feedback is assumed. An optional third argument can be used to handle the positive feedback case.	Control System
find	**find** returns the indices and values of the nonzero elements of its argument.	MATLAB
interp1	Given two vectors x and y that define the function $y(x)$ and a value x_1, **interp1** returns the interpolated value $y_1 = y(x_1)$.	MATLAB
logspace	The function **logspace** generates vectors whose elements are logarithmically spaced.	MATLAB
margin	Given a model in TF or SS form, **margin** returns the gain and phase margins and the crossover frequencies. When the output variables are omitted, it generates a Bode plot, with the margins and crossover frequencies indicated on the plot.	Control System
reshape	Given a multidimensional array, **reshape** can be used to change its dimensions.	MATLAB
roots	Given a row vector containing the coefficients of a polynomial $P(s)$, **roots** returns the solutions of $P(s) = 0$.	MATLAB
tf	Given numerator and denominator polynomials, **tf** creates the system model as a TF object. The command also converts zero-pole-gain or state-space models to TF form.	Control System

CHAPTER 10

State-Space Design Methods

PREVIEW

Compared to an input-output transfer function model, a state-space model also contains information on the internal states. Time-domain control design methods using state-space models are largely based on utilizing the internal states as part of the design scheme. A primary design technique is pole placement, in which the closed-loop system poles are placed at specific locations in the s-plane. If all the states are available for feedback control, then the design reduces to the computation of a static feedback-gain matrix. If some of the states are not measurable, then a state estimator or observer can be constructed. The separation principle allows for the independent computation of the full-state-feedback gain-matrix and the state-estimator gain-matrix. In this chapter, we will illustrate these design methods. To complete the overall picture of the design process, we will also discuss the determination of system controllability and observability.

CONTROLLABILITY

Consider the state-space model

$$\dot{\mathbf{x}}(t) = \mathbf{A}\mathbf{x}(t) + \mathbf{B}u(t), \qquad \mathbf{y}(t) = \mathbf{C}\mathbf{x}(t) \tag{10.1}$$

where $\mathbf{x}(t)$ is the n-vector of state variables, $u(t)$ is the control variable, and $\mathbf{y}(t)$ is the p-vector of output variables. In this chapter we will consider only single-input systems but allow for multiple outputs. In addition, the **D** matrix that relates the input $u(t)$ to $\mathbf{y}(t)$ is assumed to be zero and is not shown in (10.1). However, MATLAB commands that operate on state matrices will require it as an argument.

The system (10.1) is said to be controllable if there exists a constant gain matrix **F** such that the feedback control law

$$u(t) = -\mathbf{F}\mathbf{x}(t) \tag{10.2}$$

will place the poles of the closed-loop system—that is, the eigenvalues of $\mathbf{A} - \mathbf{BF}$—at any arbitrary locations. A test for controllability is that rank $(C) = n$ where C is the controllability matrix defined as

$$C = [\mathbf{B} \quad \mathbf{AB} \quad \cdots \quad \mathbf{A}^{n-1}\mathbf{B}]$$

If rank $(C) = r < n$, then only r eigenvalues of $\mathbf{A} - \mathbf{BF}$ can be arbitrarily assigned.

The following example illustrates the use of the Control System Toolbox function `ctrb` to generate C and to determine how many eigenvalues are controllable.

EXAMPLE 10.1 *Controllability Matrix*

Find the controllability matrix C for a system in SS form described by (10.1) where

$$\mathbf{A} = \begin{bmatrix} -2.0 & -2.5 & -0.5 \\ 1 & 0 & 0 \\ 0 & 1 & 0 \end{bmatrix}, \qquad \mathbf{B} = \begin{bmatrix} 1 \\ 0 \\ 0 \end{bmatrix}$$

and determine whether the system is controllable.

Solution

Using the MATLAB commands in Script 10.1, we obtain `Co` as

$$C = \begin{bmatrix} 1 & -2 & 1.5 \\ 0 & 1 & -2 \\ 0 & 0 & 1 \end{bmatrix}$$

Because `c_eig`, the rank of `Co`, is 3, the system is controllable.

MATLAB Script

```
% Script 10.1:  determine system controllability
A = [-2 -2.5 -0.5; 1 0 0; 0 1 0]     % state matrix
B = [ 1; 0; 0]                       % input matrix
Co = ctrb(A,B)                       % controllability matrix
c_eig = rank(Co)                     % number of controllable eigenvalues
```

Comment: The state-space model in Example 10.1 is in the controller canonical form. It can be shown that its controllability matrix C is always upper triangular with one's on the diagonal. Hence, C has full rank, and the system is always controllable.

REINFORCEMENT PROBLEMS

For each of the following problems, given the **A** and **B** matrices of a state-space model, calculate the controllability matrix C and test it for controllability.

P10.1 Third-order system.

$$\mathbf{A} = \begin{bmatrix} -4 & 1 & 2 \\ 1 & -5 & 3 \\ 2 & 0 & -6 \end{bmatrix} \quad \text{and} \quad \mathbf{B} = \begin{bmatrix} 1 \\ 0.5 \\ 2 \end{bmatrix}$$

P10.2 Fourth-order system in modal form.

$$\mathbf{A} = \begin{bmatrix} -2 & 0 & 0 & 0 \\ 0 & -4 & 0 & 0 \\ 0 & 0 & -5 & 0 \\ 0 & 0 & 0 & 0 \end{bmatrix} \quad \text{and} \quad \mathbf{B} = \begin{bmatrix} 1 \\ 0 \\ 2 \\ 10 \end{bmatrix}$$

Given a system in the modal form, any distinct eigenvalue in **A** corresponding to a zero entry in the **B** matrix is not controllable.

POLE PLACEMENT

Because system transient behaviors due to initial conditions and reference inputs depend directly on the poles, the objective of a pole-placement design is to apply feedback such that the system transients will decay in an acceptable period of time. When a system is controllable and all its states are available for feedback, a full-state feedback control (10.2) can be applied to place the poles of the closed-loop system at arbitrary locations in the s-plane.

The algorithm commonly cited in control textbooks for pole-placement design of single-input systems is the Ackermann formula, which is available in the Control System Toolbox as the function `acker`. However, the algorithm involves the inversion of the controllability matrix, which can be ill-conditioned for high-order systems. The Control System Toolbox function `place` utilizes a more robust pole-placement algorithm which is also applicable to multi-input systems and is recommended for general use, even for single-input systems. The use of `place` is illustrated in the next example.

EXAMPLE 10.2
Pole Placement by Full-State Feedback

Find the control gain **F** such that when the system given in Example 10.1 is controlled by (10.2), the closed-loop poles are at $s = -1, -2$, and -3. Verify the control gain by finding the eigenvalues of $\mathbf{A} - \mathbf{BF}$.

Solution

To use the `place` command, we must specify the matrices **A** and **B** and the closed-loop poles in a row vector $\mathbf{p} = [-1 \quad -2 \quad -3]$. (Note that **p** can also be a column vector.) The commands in Script 10.2 will find the required control gain and verify the result. The computation shows that

$$\mathbf{F} = [4 \quad 8.5 \quad 5.5]$$

Note that the closed-loop system matrix

$$\mathbf{A}_{\text{cl}} = \mathbf{A} - \mathbf{BF} = \begin{bmatrix} -6 & -11 & -6 \\ 1 & 0 & 0 \\ 0 & 1 & 0 \end{bmatrix}$$

is again in the controller form. The computation of the eigenvalues of $\mathbf{A}_{\text{cl}}$ indicates that **F** has been computed accurately. For high-order systems, it is a good idea to check the eigenvalues of the closed-loop system to verify the accuracy of the closed-loop poles.

MATLAB Script

```
% Script 10.2:  pole-placement design
A = [-2 -2.5 -0.5; 1 0 0; 0 1 0]   % state matrix
B = [ 1; 0; 0]                      % input matrix
p = [-1 -2 -3]                      % desired pole locations
F = place(A,B,p)                    % control gain
A_cl = A-B*F                        % closed-loop system matrix
eig(A_cl)                           % check closed-loop eigenvalues
```

Although controllability allows unlimited ability to shift the poles of a closed-loop system, the reality is that the control action of an actuator is always bounded. An extremely large controller gain that cannot be fully exercised by an actuator because of saturation may even result in an unstable closed-loop system. As a result, a practical pole-placement design must ensure that if a sufficiently large control gain is used, the saturation of the actuator would not cause undesirable system behavior. This design consideration encourages the shifting of the closed-loop poles to desirable locations not far from the open-loop poles. To measure the magnitude of a gain vector **F**, we use the 2-norm, which is defined as

$$\|\mathbf{F}\|_2 = \left(\sum_{i=1}^{n} f_i^2 \right)^{1/2}$$

where f_i, $i = 1, \ldots, n$, are the entries of **F**. The norm of **F** can be computed using the MATLAB command `norm(F)`. The following example illustrates a low-gain design.

EXAMPLE 10.3
Pole Placement (Low Gain)

For the system given in Example 10.1, shift the real pole from $s = -0.241$ to -0.5 and the complex poles from $s = -0.880 \pm j1.14$ to $-1.14 \pm j1.14$. This specification allows the transients due to the real pole to decay twice as fast and the transients due to the complex poles to be critically damped. Find the control gain **F** to achieve the closed-loop pole placement. Compare the magnitude of **F** to that obtained in Example 10.2.

Solution

We use the commands in Script 10.2 for Example 10.2 with

$$\mathbf{p} = [-0.5 \quad -1.14 + j1.14 \quad -1.14 - j1.14]$$

to find the desired control gain

$$\mathbf{F} = [0.780 \quad 1.24 \quad 0.800]$$

The command `norm(F)` yields a value of 1.67. In Example 10.2, the poles were shifted farther to the left in the s-plane, resulting in **F** having a significantly higher norm of 10.89.

REINFORCEMENT PROBLEMS

For each of the following problems, find the feedback gain matrix **F** to place the poles of the closed-loop system at the specified locations. Verify that the closed-loop pole locations are what you requested.

P10.3 Third-order system. Place the poles of the system in Problem 10.1 at $s = -4$, 8, and -10.

P10.4 Fourth-order modal system. Place the poles of the system

$$\mathbf{A} = \begin{bmatrix} -2 & 0 & 0 & 0 \\ 0 & -4 & 0 & 0 \\ 0 & 0 & -5 & 0 \\ 0 & 0 & 0 & 0 \end{bmatrix} \quad \text{and} \quad \mathbf{B} = \begin{bmatrix} 1 \\ 1 \\ 2 \\ 10 \end{bmatrix}$$

at $s = -1$, -2.5, -4.5, and -5.5.

P10.5 Lightly damped modes. Place the poles of the system

$$\mathbf{A} = \begin{bmatrix} -0.1 & 5 & 0.1 \\ -5 & -0.1 & 5 \\ 0 & 0 & -10 \end{bmatrix} \quad \text{and} \quad \mathbf{B} = \begin{bmatrix} 0 \\ 0 \\ 10 \end{bmatrix}$$

at (i) $s = -1 \pm j5$ and -10, and (ii) $s = -50, -60$, and -70. Find and compare the norms of $\mathbf{F}$.

OBSERVABILITY

Consider again the state-space model (10.1)

$$\dot{\mathbf{x}}(t) = \mathbf{A}\mathbf{x}(t) + \mathbf{B}u(t), \qquad \mathbf{y}(t) = \mathbf{C}\mathbf{x}(t) \tag{10.3}$$

The system described by (10.3) is observable if there exists a constant estimator gain matrix $\mathbf{L}$ such that the eigenvalues of $\mathbf{A} - \mathbf{LC}$ can be assigned to any arbitrary locations. A test for observability is that rank $(\mathcal{O}) = n$ where $\mathcal{O}$ is the observability matrix defined as

$$\mathcal{O} = \begin{bmatrix} \mathbf{C} \\ \mathbf{CA} \\ \vdots \\ \mathbf{CA}^{n-1} \end{bmatrix}$$

If rank $(\mathcal{O}) = r < n$, then only r eigenvalues of $\mathbf{A} - \mathbf{LC}$ can be arbitrarily assigned.

The following example illustrates the use of the Control System Toolbox function `obsv` to generate $\mathcal{O}$ and to determine the number of eigenvalues of $\mathbf{A} - \mathbf{LC}$ that can be assigned.

EXAMPLE 10.4
Observability Matrix

Find the observability matrix $\mathcal{O}$ for system (10.3) where

$$\mathbf{A} = \begin{bmatrix} -2.0 & -2.5 & -0.5 \\ 1 & 0 & 0 \\ 0 & 1 & 0 \end{bmatrix} \quad \text{and} \quad \mathbf{C} = [1 \quad 4 \quad 3.5]$$

and determine whether the system is observable.

Solution

Using the commands in Script 10.4, we obtain `Ob` as

$$\mathcal{O} = \begin{bmatrix} 1 & 4 & 3.5 \\ 2 & 1 & -0.5 \\ -3 & -5.5 & -1 \end{bmatrix}$$

Since `o_eig`, the rank of `Ob`, is 3, the system is observable.

MATLAB Script

```
% Script 10.4:  determine system observability
A = [-2 -2.5 -0.5; 1 0 0; 0 1 0]  % state matrix
C = [1 4 3.5]                     % output matrix
Ob = obsv(A,C)                    % observability matrix
o_eig = rank(Ob)                  % number of observable eigenvalues
```

REINFORCEMENT PROBLEMS

For each of the following problems, given the **A** and **C** matrices of a state-space model, calculate the observability matrix $\mathcal{O}$ and test it for observability.

P10.6 Third-order system.

$$\mathbf{A} = \begin{bmatrix} -4 & 1 & 2 \\ 1 & -5 & 3 \\ 2 & 0 & -6 \end{bmatrix} \quad \text{and} \quad \mathbf{C} = [0 \quad 1 \quad 0]$$

P10.7 Fourth-order modal system.

$$\mathbf{A} = \begin{bmatrix} -2 & 0 & 0 & 0 \\ 0 & -4 & 0 & 0 \\ 0 & 0 & -5 & 0 \\ 0 & 0 & 0 & 0 \end{bmatrix} \quad \text{and} \quad \mathbf{C} = [1 \quad 2 \quad 0 \quad 1]$$

Given a system in modal form, any distinct eigenvalue in **A** corresponding to a zero entry in the **C** matrix is not observable.

P10.8 Observer form.

$$\mathbf{A} = \begin{bmatrix} -2 & 1 & 0 & 0 \\ -3 & 0 & 1 & 0 \\ -5 & 0 & 0 & 1 \\ -10 & 0 & 0 & 0 \end{bmatrix} \quad \text{and} \quad \mathbf{C} = [1 \quad 0 \quad 0 \quad 0]$$

This system is in the observer canonical form. Its observability matrix $\mathcal{O}$ is lower triangular with one's on the diagonal. Thus $\mathcal{O}$ has full rank, and the system is always observable.

OBSERVER DESIGN

If a system described by (10.3) is completely observable, a state estimator, also called an observer, can be built using the input $u(t)$ and the output $\mathbf{y}(t)$ to estimate the state variable $\mathbf{x}(t)$. The equations of the observer, with $\mathbf{D} = \mathbf{0}$, are

$$\dot{\hat{\mathbf{x}}}(t) = \mathbf{A}\hat{\mathbf{x}}(t) + \mathbf{B}u(t) + \mathbf{L}[\mathbf{y}(t) - \hat{\mathbf{y}}(t)], \qquad \hat{\mathbf{y}}(t) = \mathbf{C}\hat{\mathbf{x}}(t) \qquad (10.4)$$

where $\hat{\mathbf{x}}(t)$ is the observer state to provide an estimate of $\mathbf{x}(t)$. Equation (10.4) can be rewritten as

$$\dot{\hat{\mathbf{x}}}(t) = (\mathbf{A} - \mathbf{LC})\hat{\mathbf{x}}(t) + \mathbf{B}u(t) + \mathbf{L}\mathbf{y}(t)$$

If all the eigenvalues of $\mathbf{A} - \mathbf{LC}$ are in the open left half-plane, then after the initial system transient has decayed, $\hat{\mathbf{x}}(t)$ will follow $\mathbf{x}(t)$. This is known as asymptotic tracking. Transposing $\mathbf{A} - \mathbf{LC}$, we note that the resulting matrix $\mathbf{A}^T - \mathbf{C}^T\mathbf{L}^T$ is like the full-state feedback matrix $\mathbf{A} - \mathbf{BF}$ except $\mathbf{A}^T$ and $\mathbf{C}^T$

replace **A** and **B**, respectively, and the design gain is $\mathbf{L}^T$ instead of **F**. Thus the command `place` can be used for designing the observer gain **L**.

In observer design, it is logical to require the state estimation error $\mathbf{x}(t) - \hat{\mathbf{x}}(t)$ to decay faster than the system transients. This is achieved by using a sufficiently large gain **L** to place the observer poles to the left of the system poles. However, **L** should not be excessive because it would result in too wide a bandwidth for the observer, which is undesirable because of the presence of noise or unmodeled high frequency dynamics. In the following example, we will use the `place` command to design an observer.

EXAMPLE 10.5
Observer Design

To provide for good tracking of the states in Example 10.4, design an observer that has its poles at $s = -3$, -5, and -7. Find the required observer gain **L** and verify the solution. Then simulate the observer when the input is zero and the initial conditions are $\mathbf{x}(0) = [1 \quad -0.75 \quad 0.4]^T$ and $\hat{\mathbf{x}}(0) = \mathbf{0}$. Draw $\mathbf{x}(t)$ and $\hat{\mathbf{x}}(t)$ on the same plot, and $\mathbf{y}(t)$ and $\hat{\mathbf{y}}(t)$ on a separate plot.

Solution

In this problem, we do not need the **B** and **D** matrices of the state-space model, since the input is zero. However, these two matrices are required as arguments in the `lsim` command. Thus we set **B** as a zero vector and **D** as a zero scalar. After using the `place` command to compute the required gain matrix, we transpose the resulting gain matrix to obtain **L**. To find $\hat{\mathbf{x}}(t)$ and $\hat{\mathbf{y}}(t)$, we first need to simulate the system with the initial condition $\mathbf{x}(0)$ using `lsim` and then take the resulting output $\mathbf{y}(t)$ as the input to the observer and simulate the observer using `lsim` again. The commands in Script 10.5 will perform these functions.

MATLAB Script

```
% Script 10.5: observer design
A = [-2 -2.5 -0.5; 1 0 0; 0 1 0]  % state matrix
B = [ 0; 0; 0]                    % input matrix set to zero
C = [1 4 3.5]                     % output matrix
D = 0                             % throughput matrix
p = [-3 -5 -7]                    % desired observer pole locations
L = place(A',C',p)'               % estimator gain
A_ob = A-L*C                      % observer system matrix
eig(A_ob)                         % check eigenvalues
x0 = [1; -0.75; 0.4]              % initial conditions
t = [0:0.04:4]';                  % time array
u = 0*t;                          % zero input
G = ss(A,B,C,D)                   % build system as LTI object
[y,t,x] = lsim(G,u,t,x0);         % simulate system to get y(t)
G_ob = ss(A_ob,L,C,D)             % build observer as LTI object
[y_hat,t,x_hat] = lsim(G_ob,y,t); % simulate observer with zero ICs
plot(t,x_hat,t,x,'--')            % figure 10.1(a)
plot(t,y_hat,t,y,'--')            % figure 10.1(b)
```

The computation shows that

$$\mathbf{L} = [35.23 \quad -19.82 \quad 16.30]^T \quad \text{and} \quad \mathbf{A} - \mathbf{LC} = \begin{bmatrix} -37.2 & -143.4 & -123.8 \\ 20.8 & 79.3 & 69.4 \\ -16.3 & -64.2 & -57.0 \end{bmatrix}$$

The eigenvalues of $\mathbf{A} - \mathbf{LC}$ are found to be at $s = -3, -5$, and -7. Figure 10.1(a) shows the time response of $\mathbf{x}(t)$ and $\hat{\mathbf{x}}(t)$. Note that initially, all the observer states in $\hat{\mathbf{x}}(t)$ move in directions opposite to the system states in $\mathbf{x}(t)$. However, after the initial transients have decayed in about 2 s, $\hat{\mathbf{x}}(t)$ tracks $\mathbf{x}(t)$. The same tracking property also applies to the output variable $\hat{\mathbf{y}}(t)$, as shown in Figure 10.1(b). In contrast to the observer states, the output variable $\hat{\mathbf{y}}(t)$ initially moves in the direction of $\mathbf{y}(t)$.

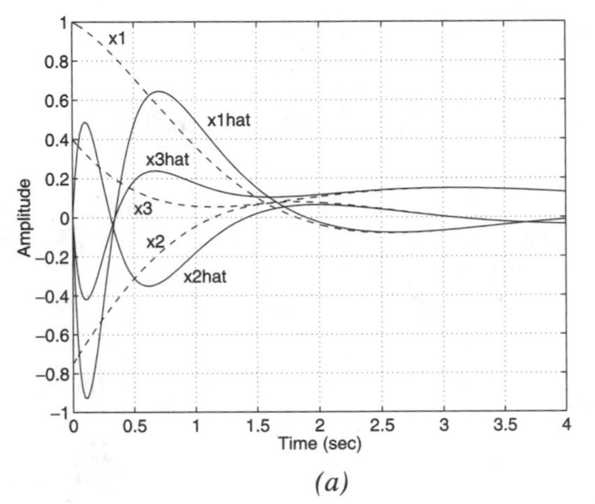

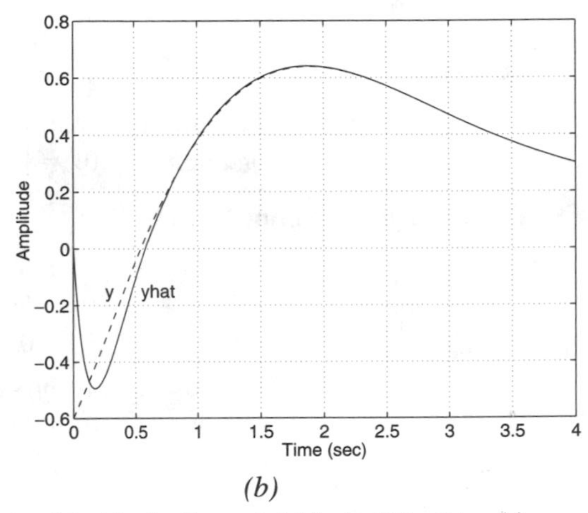

(a) (b)

FIGURE 10.1 *Observer response for Example 10.5 (a)* $\mathbf{x}(t)$ *(dashed) and* $\hat{\mathbf{x}}(t)$ *(solid) (b)* $\mathbf{y}(t)$ *(dashed) and* $\hat{\mathbf{y}}(t)$ *(solid)*

WHAT IF? Suppose that in Example 10.5 we desire the observer poles to be at $s = -6, -10$, and -14. Find the observer response and the norm of $\mathbf{L}$, and compare them to those obtained in Example 10.5. You will find that the new poles result in an $\mathbf{L}$ with a larger norm and larger initial transients in $\hat{\mathbf{x}}(t)$ but faster tracking of the system state $\mathbf{x}(t)$. ■

REINFORCEMENT PROBLEMS

In the following problems, given the matrices $\mathbf{A}$ and $\mathbf{C}$ of a state-space model, design an observer by using the gain $\mathbf{L}$ to place the poles of $\mathbf{A} - \mathbf{LC}$ at the specified locations. Simulate the observer given a zero input and the initial condition $\mathbf{x}(0)$. Assume $\hat{\mathbf{x}}(0) = \mathbf{0}$, and let $\mathbf{B}$ and $\mathbf{D}$ be zero. Make a single plot of $\mathbf{x}(t)$ and $\hat{\mathbf{x}}(t)$ and a second single plot of $\mathbf{y}(t)$ and $\hat{\mathbf{y}}(t)$.

P10.9 Third-order system.

$$\mathbf{A} = \begin{bmatrix} -4 & 1 & 2 \\ 1 & -5 & 3 \\ 2 & 0 & -6 \end{bmatrix} \quad \text{and} \quad \mathbf{C} = [0 \quad 1 \quad 0]$$

Place the eigenvalues of $\mathbf{A} - \mathbf{LC}$ at $s = -8,\ -16$, and -20. The initial condition is $\mathbf{x}(0) = [1 \quad 0.5 \quad -0.25]^T$.

P10.10 Fourth-order modal system.

$$\mathbf{A} = \begin{bmatrix} -2 & 0 & 0 & 0 \\ 0 & -4 & 0 & 0 \\ 0 & 0 & -5 & 0 \\ 0 & 0 & 0 & 0 \end{bmatrix} \quad \text{and} \quad \mathbf{C} = [1 \quad 2 \quad -1 \quad 1]$$

Place the eigenvalues of $\mathbf{A} - \mathbf{LC}$ at $s = -1.5, -3, -5.5$, and -6. The initial condition is $\mathbf{x}(0) = [1 \quad 2 \quad 0 \quad 1]^T$.

P10.11 Lightly damped modes.

$$\mathbf{A} = \begin{bmatrix} -0.1 & 5 & 0.1 \\ -5 & -0.1 & 5 \\ 0 & 0 & -10 \end{bmatrix} \quad \text{and} \quad \mathbf{C} = [1 \quad 0 \quad 0]$$

Place the eigenvalues of $\mathbf{A} - \mathbf{LC}$ at $s = -2,\ -4$, and -15. The initial condition is $\mathbf{x}(0) = [1 \quad 0 \quad 0]^T$.

OBSERVER-CONTROLLER DESIGN

The ability to assign the closed-loop poles to any location in the s-plane offers a strong motivation for using the full-state feedback gain as the controller. However, in realistic systems, not all the states are measured because of physical or economical reasons. On the other hand, the observer allows the reconstruction of the states from the outputs of the system, and hence, can be used to implement the full-state feedback control.

A key concept in the design of an observer-controller is the *separation principle* in which the full-state feedback gain $\mathbf{F}$ and the observer gain $\mathbf{L}$ are obtained independently. From these gains, an observer-controller can be constructed where the observer part provides the state estimate $\hat{\mathbf{x}}(t)$ and the full-state feedback control law is implemented as $u(t) = -\mathbf{F}\hat{\mathbf{x}}(t)$.

Following the controller designs formulated in Chapters 8 and 9, we will perform a control design for the feedback system shown in Figure 10.2, where $\mathbf{r}(t)$ is the reference input. For the single-input, multi-output systems considered in this chapter, $\mathbf{r}(t)$ is a vector having the same dimension as $\mathbf{y}(t)$, namely, $p \times 1$. The $p \times p$ gain matrix $\mathbf{N}$ is introduced as a normalization constant to ensure zero steady-state error for step inputs. This is done because, in a state-space design, the gains are selected only to shift the poles, without regard to the steady-state error.

The state-space equations of the observer-controller, with $\mathbf{D} = \mathbf{0}$, are

$$\dot{\hat{\mathbf{x}}}(t) = (\mathbf{A} - \mathbf{BF} - \mathbf{LC})\hat{\mathbf{x}}(t) - \mathbf{L}[\mathbf{Nr}(t) - \mathbf{y}(t)], \qquad u(t) = -\mathbf{F}\hat{\mathbf{x}}(t) \quad (10.5)$$

where $\hat{\mathbf{x}}$ is the state vector, $\mathbf{Nr}(t) - \mathbf{y}(t)$ is the input to the controller, and $u(t)$ is the scalar output of the controller. Thus in the state-space model (10.5), the system matrix is $\mathbf{A} - \mathbf{BF} - \mathbf{LC}$, the input matrix is $-\mathbf{L}$, and the output matrix is $-\mathbf{F}$. The closed-loop system in Figure 10.2 is of order $2n$, where n is the order of the plant as well as the observer-controller. The poles of the closed-loop system are the union of the eigenvalues of $\mathbf{A} - \mathbf{BF}$ and $\mathbf{A} - \mathbf{LC}$.

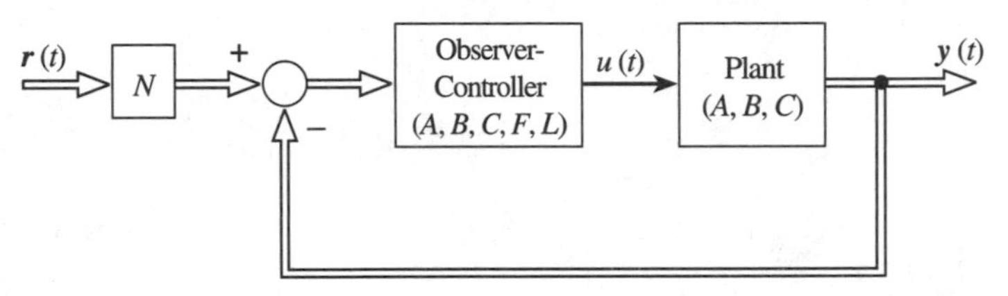

FIGURE 10.2 *Feedback system with observer-controller*

The following example illustrates the design of a controller based on (10.5).

EXAMPLE 10.6 ***Observer-Controller Design***

Combine the design results of the pole-placement controller in Example 10.2 and the observer in Example 10.5 to form the observer-controller (10.5) for the feedback system in Figure 10.2. Determine the poles and zeros of the transfer function of the observer-controller, and plot its frequency response. Form the state-space model of the closed-loop system and compute the closed-loop eigenvalues. Plot the state response of the closed-loop system due to the initial conditions $\mathbf{x}(0) = [1 \quad -0.75 \quad 0.4]^T$ and $\hat{\mathbf{x}}(0) = \mathbf{0}$, with the reference $r(t) = 0$. Then plot the step-reference response of the closed-loop system with zero initial conditions.

Solution

We begin, as shown in Part (a) of Script 10.6, by using a number of commands from Scripts 10.2 and 10.5 to obtain the full-state feedback gain $\mathbf{F}$ and the observer gain $\mathbf{L}$.

MATLAB Script

```
% Script 10.6:  observer-controller design
%-- (a) find control gain F and observer gain L
A = [-2 -2.5 -0.5; 1 0 0; 0 1 0]          % plant state matrices
B = [ 1; 0; 0], C = [1 4 3.5], D = 0
Gp = ss(A,B,C,D)                           % build plant as LTI object
p_s = [-1 -2 -3]                           % desired system poles
F = place(A,B,p_s)                         % control gain
p_o = [-3 -5 -7]                           % desired observer poles
L = place(A',C',p_o)'                      % estimator gain
%-- (b) build observer-controller according to (10.5)
Af = A-B*F-L*C                             % obsv-cont system matrix
Gc = ss(A_f,-L,-F,0)                       % build obsv-cont as LTI object
f_poles = pole(Gc)                         % obsv-cont poles
f_zeros = tzero(Gc)                        % obsv-cont zeros
bode(Gc)                                   % freq resp for figure 10.3
GpGc = Gp*Gc                               % controller & plant cascade
Gcl = feedback(GpGc,1,-1)                  % CL system; DC gain ≠ 1
cl_loop_poles = pole(Gcl)                  % CL system poles
%-- (c) pre-gain for unity DC gain
lfg = dcgain(Gcl)                          % lower frequency gain
N = 1/lfg                                  % normalization constant
%-- incorporate N in series with closed-loop system transfer function
T_ref = N*Gcl;
t = [0:0.02:4]';                           % column vector of time
r = 0*t;                                   % zero reference input
z0 = [1 -0.75 0.4 0 0 0]'                  % initial condition vector
[y,t,z] = lsim(T_ref,r,t,z0);              % IC simulation
plot(t,z(:,1:3),'--',t,z(:,4:6))           % state response in figure 10.4a
step(T_ref), grid                          % step response for figure 10.4b
[ys,t,z] = step(T_ref);                    % step response to get ys values
[Mo,tp,tr,ts,ess] = tstats(t,ys,1)         % check time-domain performance
```

From Part (b) of Script 10.6, the system matrix of the controller is found to be

$$\mathbf{A} - \mathbf{BF} - \mathbf{LC} = \begin{bmatrix} -41.23 & -151.93 & -129.31 \\ 20.82 & 79.27 & 69.36 \\ -16.30 & -64.18 & -57.04 \end{bmatrix}$$

The controller poles are at $s = -1.73$, -2.71, and -14.56, and the controller zeros are at $s = -1.60 \pm j0.435$. In general, the controller poles and zeros have no simple relationship to either the desired system poles or the observer poles. The Control System Toolbox also has a command named `reg` which can be used to design the observer-controller by entering `reg(Gp,F,-L)`.

The frequency response of the observer-controller is shown in Figure 10.3. At low frequencies the controller has a gain of 2.49 (7.9 dB). Between the

FIGURE 10.3 *Controller frequency response for Example 10.6*

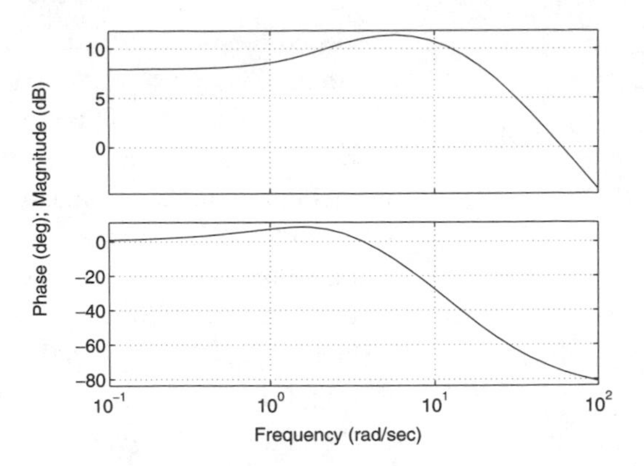

frequency range of 0.1 to 3 rad/s, it has a slight phase lead, and between 1 and 5 rad/s, the gain shows a small increase. These effects are both due to the complex zeros at $s = -1.60 \pm j0.435$. At higher frequencies, the controller gain rolls off at -20 dB/decade and the phase approaches $-90°$.

Denoting the state vector of the closed-loop system as $\mathbf{z}(t)$, which consists of the three states of the plant, $\mathbf{x}(t)$, and the three states of the observer, $\hat{\mathbf{x}}(t)$, we have

$$\mathbf{z}(t) = \begin{bmatrix} \mathbf{x}(t) \\ \hat{\mathbf{x}}(t) \end{bmatrix}$$

The closed-loop system matrix is

$$\mathbf{A}_{\text{cl}} = \begin{bmatrix} -2.0 & -2.5 & -0.5 & -4.0 & -8.5 & -5.5 \\ 1 & 0 & 0 & 0 & 0 & 0 \\ 0 & 1 & 0 & 0 & 0 & 0 \\ 35.2 & 140.9 & 123.3 & -41.2 & -151.5 & -129.3 \\ -19.8 & -79.3 & -69.4 & 20.8 & 79.3 & 69.4 \\ 16.3 & 65.2 & 57.0 & -16.3 & -65.2 & -57.0 \end{bmatrix}$$

which can be written in block-matrix form as

$$\mathbf{A}_{\text{cl}} = \begin{bmatrix} \mathbf{A} & -\mathbf{BF} \\ \mathbf{LC} & \mathbf{A} - \mathbf{BF} - \mathbf{LC} \end{bmatrix}$$

The six eigenvalues of $\mathbf{A}_{\text{cl}}$ are -1, -2, -3, -3, -5, and -7. The first three eigenvalues are due to the full-state feedback design, and the last three are due to the observer, in agreement with the separation principle.

In Part (c) of Script 10.6, the normalization constant is found to be $N = 1.057$. This gain is implemented as a state-space model with empty $\mathbf{A}$, $\mathbf{B}$, and $\mathbf{C}$ matrices and $\mathbf{D} = N$. After the gain is combined with the unnormalized closed-loop system by using the * operator we are ready to perform the required simulations.

Assigning the initial conditions in the 6-element column vector `z0`, we use the `lsim` command to generate the six state-variable responses shown in Figure 10.4(a). Comparing these responses to those in Figure 10.1(a), we again note that the observer-controller state variables in $\hat{\mathbf{x}}(t)$ start out in directions opposite to the states variables in $\mathbf{x}(t)$, but track them after the initial transients have decayed. Because the open-loop system poles at $s = -0.241$ and $-0.880 \pm j1.14$ have been shifted to the closed-loop system poles at $s = -1$, -2, and -3, which are farther to the left in the s-plane, the transients shown in Figure 10.4(a) decay to zero faster than those shown in Figure 10.1(a).

The step response $y_s(t)$ of the closed-loop system is obtained by using the `step` command and is shown in Figure 10.4(b). From the RPI function `tstats`, the step response is found to have a overshoot of 9%, a rise time of 0.29 s, a 2% settling time of 2.3 s, and zero steady-state error.

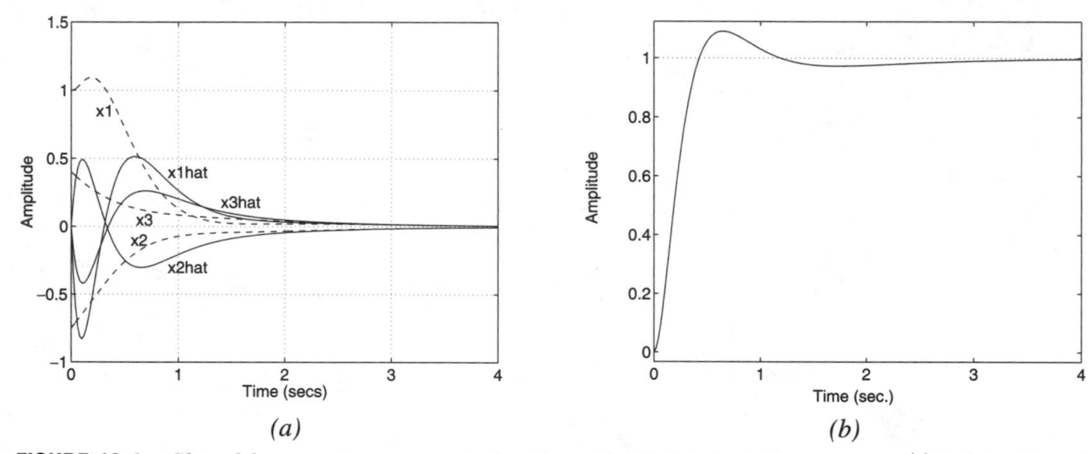

FIGURE 10.4 *Closed-loop system response for Example 10.6 (a) state response* $\mathbf{x}(t)$ *(dashed) and* $\hat{\mathbf{x}}(t)$ *(solid) (b) step response* $y_s(t)$

Comment: The Control System Toolbox also provides a command named `reg` to build the observer-controller. This command is appropriate if the observer-controller is in the feedback path instead of the forward path, as shown in Figure 10.2.

WHAT IF? Repeat Example 10.6 for some different values of the full-state feedback and observer eigenvalues to investigate the effects on the step response. See if you can reduce the 2% settling time. ■

REINFORCEMENT PROBLEMS

In each of the following problems, design an observer-controller (10.5) for the given state-space model. Use the feedback gain **F** and the observer gain **L** to place the poles of the closed-loop system at the specified locations. Determine the poles and zeros of the controller. Form the closed-loop system model and verify that the closed-loop poles are at the specified locations. Determine the gain **N** as shown in Figure 10.2 to achieve zero steady-state error for step inputs. Plot the state-variable response of the closed-loop system with $\mathbf{x}(0)$ as given in the problem statement, $\hat{\mathbf{x}}(0) = \mathbf{0}$ and $r(t) = 0$. Then plot the reference step response of the closed-loop system with zero initial conditions.

P10.12 Third-order system.

$$\mathbf{A} = \begin{bmatrix} -4 & 1 & 2 \\ 1 & -5 & 3 \\ 2 & 0 & -6 \end{bmatrix}, \quad \mathbf{B} = \begin{bmatrix} 1 \\ 0.5 \\ 2 \end{bmatrix} \quad \text{and} \quad \mathbf{C} = [0 \quad 1 \quad 0]$$

Place the full-state feedback eigenvalues at $s = -4$, -8, and -10 and the observer eigenvalues at $s = -8$, -16, and -20. The initial condition is $\mathbf{x}(0) = [1 \quad 0.5 \quad -0.25]^T$.

P10.13 Fourth-order modal system.

$$\mathbf{A} = \begin{bmatrix} -2 & 0 & 0 & 0 \\ 0 & -4 & 0 & 0 \\ 0 & 0 & -5 & 0 \\ 0 & 0 & 0 & 0 \end{bmatrix}, \quad \mathbf{B} = \begin{bmatrix} 1 \\ 1 \\ 2 \\ 10 \end{bmatrix} \quad \text{and} \quad \mathbf{C} = [1 \quad 2 \quad -1 \quad 1]$$

Place the full-state feedback eigenvalues at $s = -1$, -2.5, -4.5, and -5.5 and the observer eigenvalues at $s = -1.5$, -3, -5.5, and -6. The initial condition is $\mathbf{x}(0) = [1 \quad 2 \quad 0 \quad 1]^T$.

P10.14 Lightly damped modes.

$$\mathbf{A} = \begin{bmatrix} -0.1 & 5 & 0.1 \\ -5 & -0.1 & 5 \\ 0 & 0 & -10 \end{bmatrix}, \quad \mathbf{B} = \begin{bmatrix} 0 \\ 0 \\ 10 \end{bmatrix} \quad \text{and} \quad \mathbf{C} = [1 \quad 0 \quad 0]$$

Place the full-state feedback eigenvalues at $s = -1 \pm j5$ and -10, and the observer eigenvalues at $s = -2$, -4, and -15. The initial condition is $\mathbf{x}(0) = [1 \quad 0 \quad 0]^T$.

EXPLORATION

E10.1 Interactive analysis with MATLAB. The steps of the state-space design methods presented in this chapter are programmed in the file `ep10_1.m`. In running the file, the user can enter a state-space model, check its controllability and observability properties, assign the full-state feedback and observer eigenvalues, compute the full-state feedback gain **F** and the observer gain **L**, obtain the observer-controller system matrices and transfer function, calculate the normalization constant **N**, and simulate the state-variable response due to initial conditions and the output response due to a step reference input. You can use this file to work on any of the reinforcement problems posed in this chapter and to experiment with the effect of changing the full-state feedback and observer eigenvalues by merely entering the appropriate system matrices and the desired eigenvalues.

E10.2 Second-order plant and sensor. Develop an observer-controller for the plant and sensor that have been used in the examples of Chapters 8 and 9 by applying the state-space methods discussed in this chapter. The closed-loop system can be diagrammed as shown in Figure 10.5, where we show two outputs. The transfer functions of the plant and sensor are

$$G_p(s) = \frac{4}{(2s + 1)(0.5s + 1)} \quad \text{and} \quad H(s) = \frac{1}{0.05s + 1}$$

respectively. The output $y_p(t)$ is that of the plant, and we are interested in its response. The other output, $y_s(t)$, is that of the sensor, and it is this signal that must be used in the observer-controller calculations. Because the sensor's low-frequency gain is unity, the two signals will be identical in the steady-state.

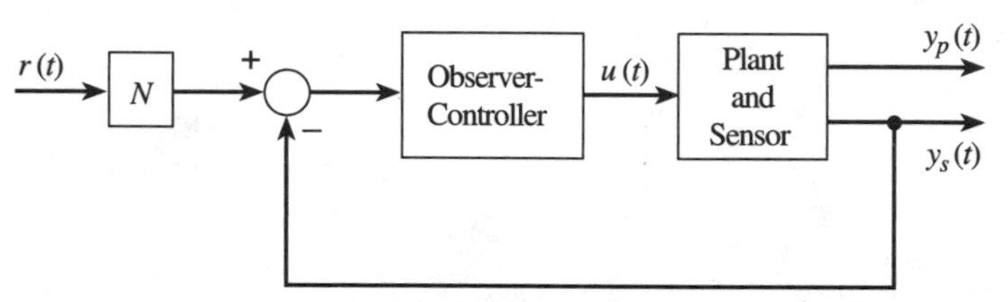

FIGURE 10.5 *Feedback system for Exploratory Problem E10.2*

Begin by building state-space models of the plant and sensor and connect them in series, with the single output being that of the sensor, $y_s(t)$. Test this third-order state-space system for controllability and observability. Then select a set of three closed-loop eigenvalues and do a pole-placement design to obtain the control gain **F**. Then select a set of eigenvalues for the observer and determine the observer gain **L**.

Modify your plant-sensor state-space model to provide the output of the plant, $y_p(t)$, and construct the observer-controller. Use the `feedback` command to connect only the sensor output to the feedback summing junction. Also use the `dcgain` command on the closed-loop system to calculate the normalization constant **N** for the reference input so the overall system will have zero steady-state error for a step input. Then simulate the response of the closed-loop system to a unit-step reference input. Your simulated step response should show both the plant and sensor outputs.

After completing your design, simulate the step response for different combinations of controller and observer eigenvalues. Try to obtain no more than 10% overshoot with a peak time of no more than 0.7 s. To gain more insight as to what is going on in this system, modify your state-space model by adding additional rows to the appropriate **C** and **D** matrices for the estimated output $\hat{y}(t)$ and the controller output $u(t)$. Having done so, use the `subplot` feature of MATLAB to plot these additional variables and try to understand their behaviors. For example, moving the observer eigenvalues farther to the left in the s-plane should reduce the observer error $y(t) - \hat{y}(t)$. Moving the controller eigenvalues to the left in the s-plane should speed up the response but at the expense of larger variations in the controller output $u(t)$, which increases the likelihood of nonlinear behavior due to actuator saturation.

COMPREHENSIVE PROBLEMS

CP10.1 Electric power generation system. Design the voltage regulator in Figure A.2 in Appendix A using the state-space methods discussed in this chapter by proceeding as follows. Run the file `epow.m` to obtain the state matrices of the power system with V_{term} as the output and compute the open-loop poles. The model has seven poles, but only the pole at $s = -0.105$ is important for voltage regulation. You can come to this conclusion by observing the open-loop step response. Thus one design strategy is to use the full-state feedback gain to shift this pole farther into the left half-plane. Similarly, in the observer, we need to track only this pole. So as a first trial, design the full-state feedback gain **F** to shift the open-loop pole at $s = -0.105$ to $s = -2$, and fix the other poles at their open-loop locations. Then design the gain **L** so that the observer will have a pole at $s = -4$ to track the pole at $s = -0.105$, with the other poles at their open-loop locations. Implement the observer-controller as $G_c(s)$ according to (10.5). Draw a Bode plot for $G_c(s)$ and simulate the step response of the closed-loop system. You will find that $G_c(s)$ behaves as a lag controller, and the step response is similar to those found in Comprehensive Problem CP8.1.

Now change the full-state feedback pole at $s = -2$ and the observer pole at $s = -4$ to some other locations, and examine the effect of the variation on the step response. You will see that as they become more negative, the rise time will become faster and the overshoot will be more significant.

You will also notice that the electromechanical mode in the observer-controller design will remain unchanged. However, this design will result in $G_c(s)$ having a notch-filter characteristic at a frequency near the electromechanical mode. The notch-filter effect is visible in the Bode plot for $G_c(s)$.

CP10.2 Satellite. Run the file `sat.m` to obtain the state-space model of the satellite and its reaction wheel. Show that this third-order system is observable but not controllable. The lack of controllability is because, in the absence of an external torque, the speed of the reaction wheel is governed by the conservation of angular momentum and cannot be controlled independently of the satellite's angular velocity. We have seen this effect in prior chapters as a pole-zero cancellation at $s = 0$ in the transfer-function model from the motor torque to the pointing angle. Fortunately, all is not lost. If you modify the state-space model by removing the wheel speed as a state variable, the resulting second-order model will be both controllable and observable, and it can be used for designing a controller for the original uncontrollable model. To accomplish this reduction, delete the third row and third column of **A**, the third element in **B**, the second row of **C**, and the second element of **D**.

Design an observer-controller for the reduced model where the input is the motor torque τ_m, in Newton-meters, and the output is the pointing angle θ, in degrees. You will have to select two controller eigenvalues and two observer eigenvalues. However, when you build the model of the closed-loop system for simulation, use the full third-order model of the satellite, with the reaction wheel included. This model will have two outputs, θ and the wheel speed Ω, in rpm. Only θ should be fed back for use in the controller calculations, but both outputs should be plotted. Try several combinations of controller and observer eigenvalues, and make note of the fact that a faster response to a change in the desired pointing angle requires greater excursions in the wheel speed. Try to achieve a 1° step response with no more than 15% overshoot, a maximum wheel-speed magnitude of 800 rpm, and a 2% settling time that does not exceed 150 s.

You should find that there will be zero steady-state pointing error for a step-reference input provided that the initial wheel speed is zero. Use the `lsim` command to simulate the response of your closed-loop system with nonzero initial wheel speeds to see what happens then.

CP10.3 Stick balancer with rigid stick. Use the file `rigid.m` to obtain the state-space model of the stick balancer with a rigid stick. The model has the single input $u(t)$, which is the voltage applied to the armature winding of the electric motor that drives the cart and the two outputs $\theta(t)$ and $x(t)$, which are the stick angle and the cart position, respectively. Design an observer-controller for this system that uses both outputs, so the observer gain matrix **L** will have four rows and two columns. Your simulations of the closed-loop system should make use of the `subplot` command to produce one plot of θ and a second plot of both x and the reference input.

Try to find controller and observer eigenvalues (four of each) that will provide a response to a 1-foot step in desired cart position that has no more than 20% overshoot, a peak time of no more than 2 s, and a maximum deviation of the stick from vertical of no more than 0.1°. You should find that the cart starts in the direction opposite to its final position to get the stick falling toward its destination. Also, simulate the response to an initial stick angle of 1°, with the observer and other plant states all starting at zero. In this case the observer must recover from an initial 1° error while trying to return the stick to vertical and the cart to its original position. You may notice that the cart goes through a substantial excursion, say on the order of 10 feet, before returning to its original position.

CP10.4 Stick balancer with flexible stick. Run the file `flex.m` to obtain the state-space model of the cart and a flexible stick. Then carry out the same analysis and design procedure described in Comprehensive Problem CP10.3. Your observer-controller will have 6 state variables, which, when added to the 6 states of the plant, will make a 12th order system. The observer gain **L** will have six rows and two columns, and the controller gain **F** will have one row and six columns. You will have to select six eigenvalues for the controller design and another six for the observer design. Simulate the response of both the measured stick angle, which is a combination of the rigid and the bending modes and the cart position. See how well you can do in terms of reducing the overshoot in the cart position and the excursion of the measured stick angle in response to a 1-foot change in the cart position.

SUMMARY

In this chapter, we used MATLAB to determine controllability and observability of state-space models, to compute the full-state feedback gain and the observer gain for eigenvalue assignment, and to investigate the performance of the observer-controller. This chapter serves as an introduction to the design of controllers using state-space methods. More advanced state-space design algorithms such as the symmetric root-locus and linear quadratic regulator design can be found in many control textbooks.

MATLAB FUNCTIONS USED

Function	*Purpose and Use*	*Toolbox*
*	Given two LTI objects, the * operator forms their series connection.	Control System
bode	Given a model in TF or SS form, **bode** returns the magnitude and phase of the frequency response. When the output variables are omitted, it generates the Bode plot directly.	Control System
ctrb	Given the **A** and **B** matrices of a system, **ctrb** returns the controllability matrix.	Control System
dcgain	Given the models as an LTI object, **dcgain** returns the steady-state gain of the system.	Control System
eig	Given a square matrix, **eig** computes its eigenvalues and eigenvectors.	MATLAB
feedback	Given the models of two systems in TF form, **feedback** returns the model of the closed-loop system, where negative feedback is assumed. An optional fifth argument can be used to handle the positive feedback case.	Control System
lsim	Given a state-space model of a continuous system, an array of input values, a vector of time points, and a vector of initial conditions, **lsim** returns the time response of the output and the state.	Control System
norm	Given a vector or a matrix, **norm** computes its norm.	MATLAB
obsv	Given the **A** and **C** matrices of a system, **obsv** returns the observability matrix.	Control System
place	Given the **A** and **B** matrices of a system, **place** returns the gain matrix **F** that places the eigenvalues of $\mathbf{A} - \mathbf{BF}$ at specified locations in the s-plane.	Control System
pole	Given an LTI object, **pole** computes the poles of the system's transfer function.	Control System

rank	Given a matrix, **rank** determines the number of linearly independent rows or columns in the matrix.	MATLAB
ss	Given a set of state-space matrices, **ss** creates the model as an SS object.	Control System
tstats	Given a step response, **tstats** finds the percent overshoot, peak time, rise time, settling time, and steady-state error.	RPI
tzero	Given a state-space model, **tzero** returns the zeros of its transfer function.	Control System

ANSWERS

P10.1 System is controllable

P10.2 Eigenvalue -4 is not controllable

P10.3 $\mathbf{F} = [-1.99 \quad 3.42 \quad 3.64]$

P10.4 $\mathbf{F} = [0.365 \quad 0.422 \quad 0.083 \quad 0.155]$

P10.5 (i) $\mathbf{F} = [-0.140 \quad 0.375 \quad 2.18]$, norm($\mathbf{F}$) = 2.22,
(ii) $\mathbf{F} = [821.7 \quad 196.3 \quad 18.98]$, norm($\mathbf{F}$) = 845

P10.6 System is observable

P10.7 Eigenvalue -5 is not observable

P10.9 $\mathbf{L} = [99.6 \quad 29.0 \quad 49.8]^T$

P10.10 $\mathbf{L} = [0.583 \quad 0.469 \quad 0.233 \quad 3.71]^T$

P10.11 $\mathbf{L} = [10.8 \quad -7.82 \quad 10.0]^T$

P10.12 controller poles: -13.3, -16.9, and -20.9; controller zeros: -6.75 and -14.63; $N = 2.01$

P10.13 controller poles: -1.42, -3.18, -5.60, and -8.30; controller zeros: -1.37, -3.37, and -4.95; $N = 1.0$

P10.14 controller poles: $-4.0 \pm j1.43$, and -14.8, controller zeros: -5.37 and -10; $N = -0.876$

Models of Practical Systems

PREVIEW

In this appendix we present simplified models of four practical plants or processes for which control systems can be developed. These models are used in the comprehensive problems throughout the book to allow the reader to apply the MATLAB commands in a quasi-realistic environment (see Table A.1 for the specific problems). As we shall see, these systems have some interesting behavior that is generally not encountered in the typical textbook problem. Commands for creating these models in both transfer-function and state-space form are contained in M-files that are available from the Brooks/Cole web site. To load a model, the reader need only enter the name of the appropriate M-file and make a selection from the menu that will appear.

Keep in mind that a *linear* model has been obtained from a *nonlinear* model at some equilibrium condition. Hence, the linear model will always have a limited range of applicability. Although we have not derived the nonlinear models in the appendix, a bit of common sense will go a long way in deciding what constitutes reasonable limits of usefulness.

APPENDIX A

TABLE A.1 *Cross-reference of practical systems and comprehensive problems*

Chapter and Topic	*Electric Power System*	*Reaction-wheel Satellite*	*Rigid Stick Balancer*	*Flexible Stick Balancer*	*Hydro-Turbine*
2. Single-block systems	CP2.1	CP2.2	CP2.3	CP2.4	CP2.5
3. Multi-block sytems	CP3.1	CP3.2,3,4	-	-	CP3.5
4. State-space models	CP4.1	CP4.2	CP4.3	CP4.4	-
5. Root locus	CP5.1	CP5.2	CP5.3	-	CP5.4
6. Frequency response	CP6.1	CP6.2	CP6.3	-	CP6.4
7. System performance	CP7.1	CP7.2	-	-	CP7.3
8. PID design	CP8.1	CP8.2,3	-	-	-
9. Lead-lag design	CP9.1	CP9.2	CP9.3	-	-
10. State-space design	CP10.1	CP10.2	CP10.3	CP10.4	-

ELECTRIC POWER SYSTEM

A single-machine infinite-bus system in (Figure A.1) is usually used as the first step in designing an excitation system control for a power plant delivering an electrical power P. The objective is to design a feedback controller with output $u(t)$ to regulate the field voltage such that the machine terminal voltage V_{term} is maintained at a desired value V_{ref}. In this example, the machine model includes subtransient effects, and the field voltage actuator is a solid state rectifier. The states for the machines are the machine rotor angle δ, the machine speed ω, and the machine direct- and quadrature-axis fluxes E_q', ψ_{1d}, E_d', and ψ_{1q}. The machine rotor angle δ is in radians, the machine speed ω in pu (per unit) normalized with respect to the synchronous speed, and the fluxes are also in pu normalized with respect to the rated voltage. The actuator is modeled with the state V_R. For a particular operating condition, the linearized system model is given by

$$\dot{\mathbf{x}} = \mathbf{A}\mathbf{x} + \mathbf{B}u(t), \qquad \mathbf{y} = \mathbf{C}\mathbf{x}$$

where

$$\mathbf{x} = [\delta \quad \omega \quad E'_q \quad \psi_{1d} \quad E'_d \quad \psi_{1q} \quad V_R]^T$$

$$\mathbf{y} = [V_{\text{term}} \quad \omega \quad P]^T$$

$$\mathbf{A} = \begin{bmatrix} 0 & 377.0 & 0 & 0 & 0 & 0 & 0 \\ -0.246 & -0.156 & -0.137 & -0.123 & -0.0124 & -0.0546 & 0 \\ 0.109 & 0.262 & -2.17 & 2.30 & -0.0171 & -0.0753 & 1.27 \\ -4.58 & 0 & 30.0 & -34.3 & 0 & 0 & 0 \\ -0.161 & 0 & 0 & 0 & -8.44 & 6.33 & 0 \\ -1.70 & 0 & 0 & 0 & 15.2 & -21.5 & 0 \\ -33.9 & -23.1 & 6.86 & -59.5 & 1.50 & 6.63 & -114.0 \end{bmatrix}$$

$$\mathbf{B} = [0 \quad 0 \quad 0 \quad 0 \quad 0 \quad 0 \quad 17.6]^T$$

$$\mathbf{C} = \begin{bmatrix} -0.123 & 1.05 & 0.230 & 0.207 & -0.105 & -0.460 & 0 \\ 0 & 1 & 0 & 0 & 0 & 0 & 0 \\ 1.42 & 0.900 & 0.787 & 0.708 & 0.0713 & 0.314 & 0 \end{bmatrix}$$

The dominant mode of the system is a pair of lightly damped poles (swing mode) corresponding to the mechanical oscillations of the machine versus the infinite bus.

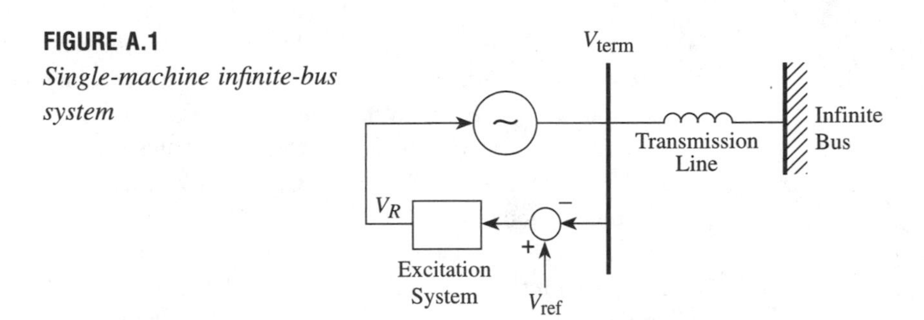

FIGURE A.1
Single-machine infinite-bus system

This system is used in several comprehensive problems for the design of a voltage control system (Figure A.2). The terminal voltage (V_{term}) is to be regulated with respect to a reference voltage. The dynamics of the sensor for measuring the terminal voltage is neglected. The gain of the voltage regulator is limited by the lightly damped mode, as will be seen in the comprehensive problems. In practice, a second feedback loop using either the speed or power output is applied to improve the damping on this lightly damped mode. In this book, we simplify this process by adding some mechanical damping to this mode in the system model.

The model of the power system is contained in the file `epow.m`, and the model with additional damping is in the file `epow2.m`. Because this is a multi-output system, the user should select from the menu whichever output is to be considered. For a state-space model, the appropriate row of the **C** matrix should be used.

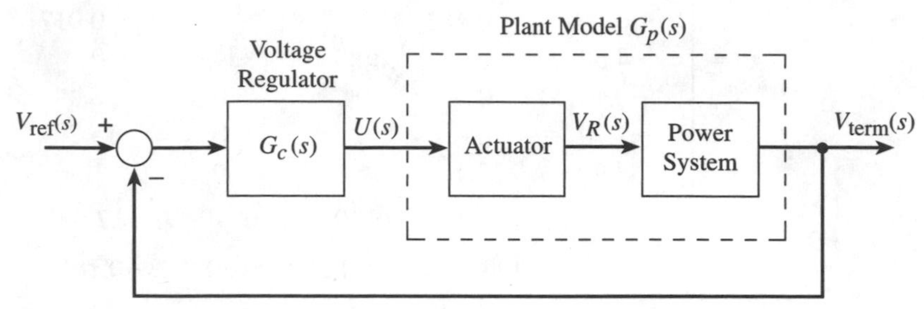

FIGURE A.2 *Excitation control system*

SATELLITE WITH REACTION WHEEL

Figure A.3 shows a satellite whose pointing angle θ is controlled by varying the speed of a reaction wheel. The reaction wheel can be thought of as an electric motor with a flywheel attached to it. The angular velocity of the reaction wheel relative to the satellite can be varied by changing the voltage applied to its armature winding. Because of the reaction torque between the satellite and the motor, there is no change in the total angular momentum of the combined satellite and reaction wheel. Thus, as the speed of the wheel relative to the satellite is varied, the angular velocity of the satellite with respect to an inertial reference frame must vary such that the total angular momentum remains constant.

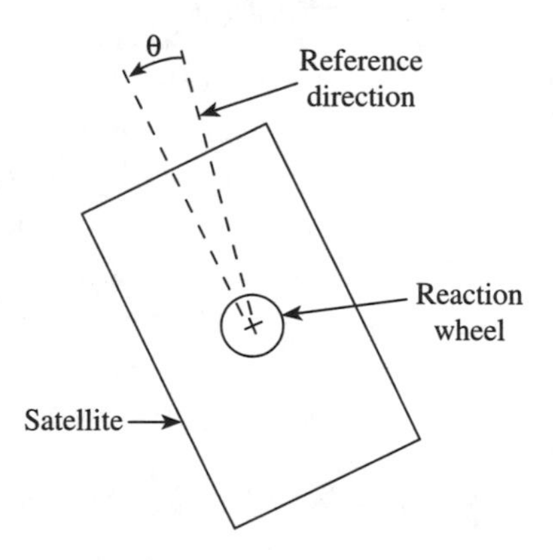

FIGURE A.3 *A satellite with a reaction wheel*

APPENDIX A

By drawing free-body diagrams for the satellite without the reaction wheel and for the reaction wheel alone and summing the moments about the axis of rotation, we can obtain the following set of state-variable equations:

$$\dot{\theta} = \omega$$

$$\dot{\omega} = \frac{B}{J_s}\Omega - \frac{1}{J_s}\tau_m(t)$$

$$\dot{\Omega} = -\frac{B}{J_{\mathrm{eq}}}\Omega + \frac{1}{J_{\mathrm{eq}}}\tau_m(t)$$

where θ is the pointing angle of the satellite, ω is the angular velocity of the satellite, and Ω is the speed of the reaction wheel relative to the satellite. The torque τ_m is developed by the motor that drives the reaction wheel. The physical parameters involved in the model are B, the viscous coefficient of friction between the reaction wheel and the satellite; J_s, the moment of inertia of the satellite (excluding the reaction wheel); J_w, the moment of inertia of the reaction wheel, and J_{eq}, the equivalent moment of inertia where

$$\frac{1}{J_{\mathrm{eq}}} = \frac{1}{J_s} + \frac{1}{J_w}$$

The model can be written in state-space form as

$$\frac{d}{dt}\begin{bmatrix} \theta \\ \omega \\ \Omega \end{bmatrix} = \begin{bmatrix} 0 & 1 & 0 \\ 0 & 0 & B/J_s \\ 0 & 0 & -B/J_{\mathrm{eq}} \end{bmatrix}\begin{bmatrix} \theta \\ \omega \\ \Omega \end{bmatrix} + \frac{1}{J_s}\begin{bmatrix} 0 \\ -1 \\ J_{\mathrm{eq}}/J_s \end{bmatrix}\tau_m(t)$$

For the parameter values $J_s = 13.6$ kg-m^2, $J_w = 13.6 \times 10^{-4}$ kg-m^2, and $B = 1.01 \times 10^{-6}$ N-m-s/rad, the numerical form of the state-space model is

$$\frac{d}{dt}\begin{bmatrix} \theta \\ \omega \\ \Omega \end{bmatrix} = \begin{bmatrix} 0 & 1 & 0 \\ 0 & 0 & 7.4265 \times 10^{-8} \\ 0 & 0 & -7.4265 \times 10^{-3} \end{bmatrix}\begin{bmatrix} \theta \\ \omega \\ \Omega \end{bmatrix} + 0.07353\begin{bmatrix} 0 \\ -1 \\ 10^5 \end{bmatrix}\tau_m(t)$$

The 1-input/3-output state-space model can be obtained in numerical form by running the file `sat.m`. This file will also generate the transfer function from the motor torque (τ_m) to the pointing angle of the satellite (θ) as a ratio of polynomials. Because the state-space model has three outputs, we must use only the appropriate rows of the **C** and **D** matrices for this calculation. For example, when the first output is involved, we use only the first rows of **C** and **D**. Regardless of which outputs are used, all of the rows and columns of **A** and **B** are included. To avoid numerically-induced problems, such as very large but finite transfer-function zeros, the RPI function `small20` is used to replace any entries in the state-space matrices that have magnitudes less than 10^{-10}.

The control system shown in Figure A.4 will be used in several comprehensive problems. The difference between the desired angle θ_{des} and the actual angle θ is the error, which is the input to the controller. The controller output $u(t)$ is the torque applied by the electric motor to the reaction wheel. The transfer function $G_1(s)$, from the motor torque to the pointing angle θ, can be obtained in TF form by running the MATLAB file `sat.m` and selecting the first entry in the menu of transfer functions. Similarly, we can get the transfer function $G_2(s)$, from the motor torque to the wheel speed Ω by selecting the second entry in the menu.

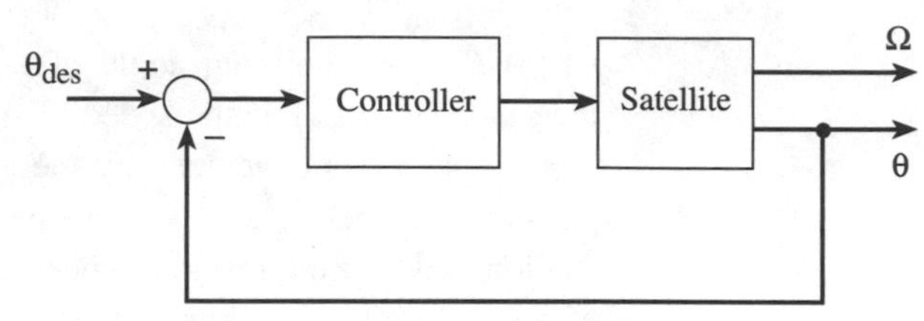

FIGURE A.4 *Closed-loop system with satellite and a controller*

FLEXIBLE STICK BALANCER

Figure A.5 shows a cart on wheels that can be moved back and forth to balance a flexible stick that is supported by a frictionless pivot at the base of the stick. The system also includes an electric motor, a cable connecting the motor to the cart, and pulleys that are not shown in the figure. There are sensors to measure the position and velocity of the cart and the angle that the base of the stick makes with the vertical, and its derivative. In order to develop a control system we want to have a linear mathematical model in state-space form.

FIGURE A.5 *Cart supporting a flexible stick*

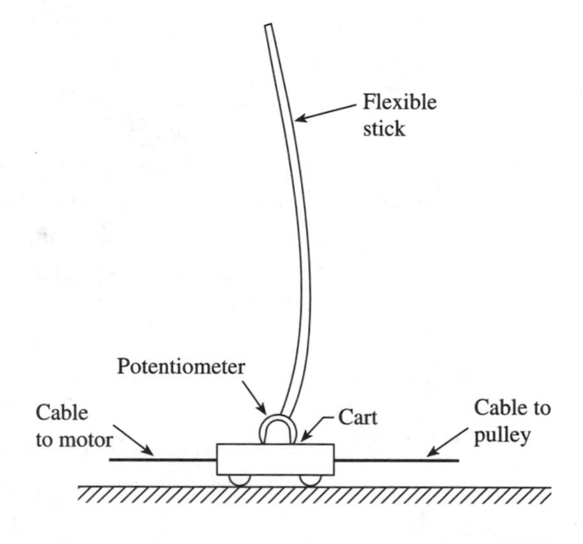

Such a model can be developed by assuming that the stick is rigid and drawing free-body diagrams of the stick, cart, electric motor, and pulleys. From these diagrams the forces and moments can be summed to obtain the equations of motion. The nonlinear terms can be neglected provided that the angle of the stick with respect to the vertical remains sufficiently small (say, no more than $\pm 10°$). The bending of the stick can be approximated by including an undamped (or very lightly damped) second-order submodel that accounts for the first bending mode of the stick. Although the determination of the frequency and shape of the first bending mode involves the development and solution of a partial differential equation, the effect on the state-space model will be to add two more state variables to the four that describe the behavior of the rigid stick, cart, and motor. Thus, we have two models: a fourth-order one for the rigid stick and a sixth-order one for the flexible stick.

The variables used to describe the system are: θ, the angle of the rigid stick relative to the vertical, in radians; ω, the derivative of that angle, in rad/s; x, the horizontal position of the cart, in feet; v, the velocity of the cart, in ft/s; u, the control signal applied to the armature of the motor, in volts; z, the amplitude of the first bending mode, which is dimensionless; and μ, the derivative of z, with units seconds^{-1}. The parameters that appear in the $\mathbf{A}$ and $\mathbf{B}$ matrices below can be expressed in terms of fundamental parameters such as dimensions and masses, but we will just give their numerical values and units later in the discussion. The interested reader can consult Franklin, Powell, and Emami-Naeini (1994) for a derivation of the model with a rigid stick.

Drawing free-body diagrams and neglecting the nonlinear terms, we can obtain the following set of coupled differential equations:

$$\dot{\omega} - p^2\theta - \beta K_p v = -K_m K_p u \qquad \text{(rotational)}$$
$$\dot{v} + \alpha\theta + \beta v = K_m u \qquad \text{(translational)}$$
$$\ddot{z} + q^2 z + \beta K_b v = K_b K_m u \qquad \text{(bending)}$$

If we select the state vector to be $[\theta\ \omega\ x\ v\ z\ \mu]^T$, the model can be written in state-space form as

$$\frac{d}{dt}\begin{bmatrix}\theta\\ \omega\\ x\\ v\\ z\\ \mu\end{bmatrix} = \begin{bmatrix}0 & 1 & 0 & 0 & 0 & 0\\ p^2 & 0 & 0 & \beta K_p & 0 & 0\\ 0 & 0 & 0 & 1 & 0 & 0\\ -\alpha & 0 & 0 & -\beta & 0 & 0\\ 0 & 0 & 0 & 0 & 0 & 1\\ 0 & 0 & 0 & -\beta K_p & -q^2 & 0\end{bmatrix}\begin{bmatrix}\theta\\ \omega\\ x\\ v\\ z\\ \mu\end{bmatrix} + \begin{bmatrix}0\\ -K_m K_p\\ 0\\ K_m\\ 0\\ K_b K_m\end{bmatrix} u(t)$$

Figure A.6(a) shows the positive senses of the variables that describe the motion of the system, associated with the rigid model of the stick. In Figure A.6(b) we see the shape of the flexible stick when the bending-mode variable z is positive. If z is negative, the stick would be bent the opposite way, with the pivot point and the node always remaining on the dashed line that represents the angle and position of the rigid mode. To get the actual position of any point on the stick at a given time, the deflection due to the bending is added to that due to the rigid mode.

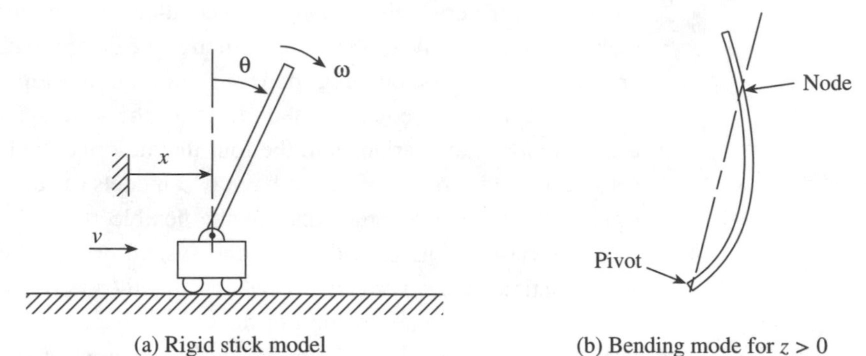

(a) Rigid stick model (b) Bending mode for $z > 0$

FIGURE A.6 *Positive senses of the variables that describe the motion of the system (a) Rigid stick model (b) Bending mode*

For control purposes, we assume that there are only two output variables that can be measured electrically. One of these is the cart position x, which is obtained from a potentiometer connected to the motor. The other signal is obtained from a potentiometer that is mounted on the cart and connected to the base of the stick. If the stick is rigid, the sensor will measure θ. However, if the bending of the stick is included, the potentiometer will measure a variable denoted by θ_m, which will be a linear combination of the rigid mode θ and the amplitude of the bending mode z. The expression for the measured angle can be written as $\theta_m = \theta + \gamma z$, where the coefficient γ involves the slope of the tangent to the bending mode curve at the base of the stick.

The file `rigid.m` develops the model when the stick is assumed to be rigid in either TF or SS form, depending on the menu selections made by the user. The file assigns parameter values that correspond to a stick six feet in length and weighing 0.2 pounds. Also included are the cart, which weighs 2 pounds, and the electric motor that results in a mechanical time constant of 10 seconds for the combined cart and motor. The values of the parameters that appear in the model are: $p = \sqrt{8}\ \text{s}^{-1}$, $\alpha = 0.15\ \text{ft/s}^2$, $\beta = 10.0$ s, $K_p = 0.25\ \text{ft}^{-1}$, $K_m = 4.64\ \text{ft/(s}^2\ \text{V)}$, and $K_b = -8.90\ \text{ft}^{-1}$.

The file `flex.m` develops the TF and SS models when the flexibility of the stick is included, with the natural frequency being $q = 20$ rad/s and no mechanical damping. A value of 0.9 is used for the parameter γ in the expression for the output of the potentiometer at the base of the stick. Both of these M-files make use of the RPI function `smal120` to eliminate very small matrix elements that can be troublesome in a numerical sense.

HYDRO-TURBINE AND PENSTOCK

The nonlinear power generation equations of the hydro-turbine and penstock system in Figure A.7 are

$$H = v^2/Q^2, \qquad \dot{v} = (g/L)(H_o - H), \qquad P = avH$$

where H is the pressure at the gate, v the water velocity, Q the effective gate opening, g the gravity constant, L the length of the penstock, H_o the reservoir water pressure, P the mechanical power generated by the turbine, and a the area of the penstock. The units of these variables are not needed in this problem. Since the power P is proportional to the water velocity v, to increase P, one has to increase the gate opening Q such that v will increase. However, since the water velocity in the penstock does not change instantaneously, if Q is increased, H will momentarily decrease. As a result, P will first decrease until the water velocity increases sufficiently. Such systems exhibiting the behavior that *it will get worse before it gets better* are quite common in engineering and economic systems.

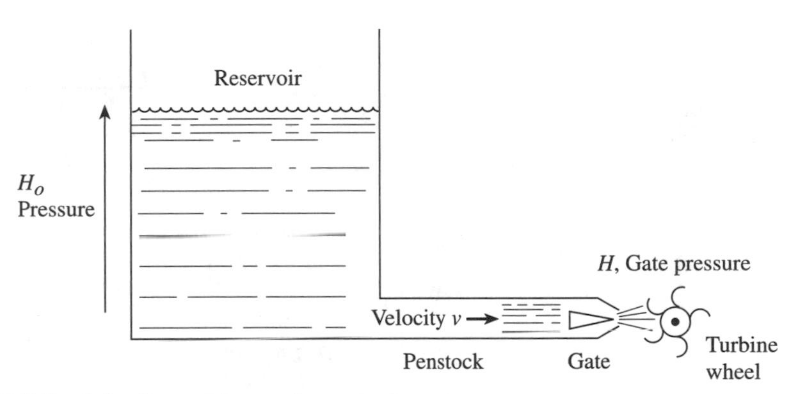

FIGURE A.7 *A hydro-turbine and penstock system*

Assuming that the nonlinear model is in an equilibrium condition with $H = H_o$ and $v = v_o$, we obtained the linearized model of the hydro-turbine as the transfer function

$$\Delta P(s) = \left(\frac{1 - sT_w}{1 + sT_w/2} \right) \Delta Q(s)$$

where $\Delta P(s)$ and $\Delta Q(s)$ are the Laplace transforms of the incremental power and gate opening, respectively, and $T_w = (v_o l)/(h_o g)$ is the water time constant. The system has a right-half-plane zero at $s = 1/T_w$.

To complete the model, we include the dynamics of the movement of the gate opening with an actuator by

$$\Delta Q(s) = \left(\frac{1}{T_Q s + 1} \right) \Delta U(s)$$

where T_Q is the actuator time constant, and $\Delta U(s)$ is the transform of the incremental actuation signal generated by the controller. Typical time constants are $T_w = 2$ s and $T_Q = 0.5$ s.

The model and data of the hydro-turbine system are contained in the file `hydro.m`. The system will be used in several comprehensive problems to analyze the effect of a right half-plane zero on the system response. It will also be used to study the generation control system in Figure A.8. For simplicity, we will drop the Δs from the incremental variables from now on. In this control system, the controller utilizes the difference between the desired power P_{des} and the actual power P to generate an output $u(t)$ to control the gate position.

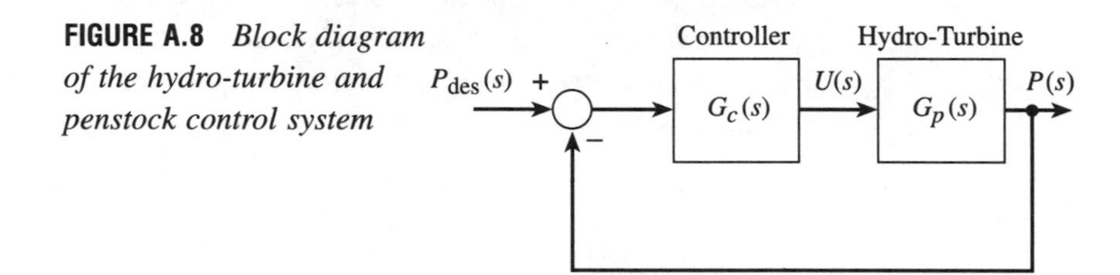

FIGURE A.8 *Block diagram of the hydro-turbine and penstock control system*

MATLAB Commands

PREVIEW

This appendix contains a summary of all the MATLAB commands, including those of the Control System Toolbox and the RPI functions, that have appeared in the summary tables at the end of each chapter. To assist in finding examples of the use of these commands, we have included a table that will serve as a comprehensive index. The pages in the text where a particular command is mentioned, the numbers of the examples, MATLAB scripts, reinforcement problems, and comprehensive problems are listed.

MATLAB FUNCTIONS USED

Function	*Purpose and Use*	*Toolbox*
*	Given two LTI objects, the * operator forms their series connection.	Control System
+	Given two LTI objects, the + operator forms their parallel connection.	Control System
angle	Given a complex number, **angle** returns the phase angle, in radians.	MATLAB
axis	**axis([xmin xmax ymin ymax])** specifies the plotting area. **axis equal** forces uniform scaling for the real and imaginary axes.	MATLAB
bode	Given a model in TF or SS form, **bode** returns the magnitude and phase of the frequency response. When the output variables are omitted, it generates the Bode plot directly.	Control System
bwcalc	Given the magnitude of the frequency response of a system and its low frequency gain, **bwcalc** computes the bandwidth.	RPI function
canon	Given a system in state-space form, **canon** returns its modal form.	Control System
conv	Given two row vectors containing the coefficients of two polynomials, **conv** returns a row vector containing the coefficients of the product of the two polynomials.	MATLAB
cpole2t	Given a complex pair of transfer-function poles, the residues at those poles, and a time vector, **cpole2t** returns the time response due to those poles.	RPI function
ctrb	Given the **A** and **B** matrices of a system, **ctrb** returns the controllability matrix.	Control System
damp	Given a model as an LTI object, **damp** calculates the natural frequencies and the damping ratios of the system poles. When invoked without output variables, a table of poles, damping ratios, and natural frequencies, is displayed.	Control System

Function	*Purpose and Use*	*Toolbox*
dcgain	Given a model as an LTI object, **dcgain** returns the steady-state gain of the system.	Control System
deconv	Given two polynomials as row vectors, **deconv** returns the quotient and remainder of the first polynomial divided by the second.	MATLAB
eig	Given a square matrix, **eig** computes its eigenvalues and eigenvectors.	MATLAB
feedback	Given the models of two systems as LTI objects, **feedback** returns the model of the closed-loop system, where negative feedback is assumed. An optional third argument can be used to handle the positive feedback case.	Control System
find	**find** returns the indices and values of the nonzero elements of its argument, which may be a logical expression.	MATLAB
findobj	Given a set of handle-graphics objects, **findobj** returns the handles of those objects having the specified property.	MATLAB
get	**get** returns the properties of an LTI object.	MATLAB
hold	When set "on," **hold** draws subsequent plots on the current set of axes.	MATLAB
impulse	Given a TF model of a continuous system, **impulse** returns the response to a unit-impulse input.	Control System
interp1	Given two vectors x and y that define the function $y(x)$ and a value x_1, **interp1** returns the interpolated value $y_1 = y(x_1)$.	MATLAB
logspace	The function **logspace** generates vectors whose elements are logarithmically spaced.	MATLAB
lsim	Given a continuous system as an LTI object, a vector of input values, a vector of time points, and possibly a set of initial conditions, **lsim** returns the time response.	Control System
margin	Given a model as an LTI object, **margin** returns the gain and phase margins and the crossover frequencies. When the output variables are	Control System

Function	Purpose and Use	Toolbox
	omitted, it generates a Bode plot with the margins and crossover frequencies indicated on the plot.	
ngrid	The function **ngrid** generates gridlines for a Nichols plot.	Control System
nichols	Given a model as an LTI object, **nichols** returns the magnitude and phase of the frequency response. When the output variables are omitted, it generates the Nichols plot directly.	Control System
norm	Given a vector or a matrix, **norm** computes its norm.	MATLAB
nyquist	Given a model as an LTI object, **nyquist** returns the real and imaginary parts of the frequency response. When the output variables are omitted, it generates the Nyquist plot directly.	Control System
obsv	Given the **A** and **C** matrices of a system, **obsv** returns the observability matrix.	Control System
place	Given the **A** and **B** matrices of a system, **place** returns the gain matrix **F** that places the eigenvalues of $\mathbf{A} - \mathbf{BF}$ at specified locations in the s-plane.	Control System
pole	Given an LTI object, **pole** computes the poles of the system's transfer function.	Control System
pzmap	Given a system model in TF or ZP form, **pzmap** produces a plot of the system's poles and zeros in the s-plane.	Control System
rank	Given a matrix, **rank** determines the number of linearly independent rows or columns in the matrix.	MATLAB
reg	Given state-feedback and estimator gains, **reg** produces an observer-based regulator as an LTI object.	Control System
reshape	Given a multidimensional array, **reshape** can be used to change its dimensions.	MATLAB
residue	Given a rational function $T(s) = N(s)/D(s)$, **residue** returns the roots of $D(s) = 0$, the partial-fraction coefficients, and any polynomial term that remains.	MATLAB
rlocfind	Given a TF or state-space description of an open-loop system, **rlocfind** allows the user to select	Control System

Function	*Purpose and Use*	*Toolbox*
	any point on the locus with the mouse and returns the value of the loop gain that will make that point be a closed-loop pole. It also returns the values of all the closed-loop poles for that gain value.	
rlocus	Given a TF or state-space description of an open-loop system, **rlocus** produces a root-locus plot that shows the locations of the closed-loop poles in the s-plane as the loop gain varies from 0 to infinity.	Control System
roots	Given a row vector containing the coefficients of a polynomial $P(s)$, **roots** returns the solutions of $P(s) = 0$.	MATLAB
rpole2t	Given a transfer-function pole, the residue at that pole, and a time vector, **rpole2t** returns the time response due to that pole.	RPI function
semilogx	The function **semilogx** generates semi-logarithmic plots, using a base 10 logarithmic scale for the x-axis and a linear scale for the y-axis.	MATLAB
set	Given a handle-graphics object, **set** assigns the value of the specified property.	Control System
ss	Given a set of state-space matrices, **ss** creates the model as an SS object.	Control System
sgrid	When viewing either a root-locus plot or a pole-zero map, **sgrid** draws contours of constant damping ratio (ζ) and natural frequency (ω_n).	Control System
small20	**small20** replaces very small matrix elements by 0	RPI function
step	Given a TF model of a continuous system, **step** returns the response to a unit-step function input.	Control System
subplot	**subplot** allows the plotting window to be divided into multiple plotting areas.	MATLAB
tf	Given numerator and denominator polynomials, **tf** creates the system model as a TF object. The command also converts zero-pole-gain or state-space models to TF form.	Control System

Function	*Purpose and Use*	*Toolbox*
tfdata	Given a TF object, **tfdata** extracts the numerator and denominator polynomials and other information about the system.	Control System
tstats	Given a step response, **tstats** finds the percent overshoot, peak time, rise time, settling time, and steady-state error.	RPI function
tzero	Given a state-space model, **tzero** returns the zeros of its transfer function.	Control System
vgain	Given a model as an LTI object, **vgain** computes the velocity error constant.	RPI function
zpk	Given a system's zeros, poles, and gain, **zpk** creates the system model as a ZPK object. The command also converts transfer-function or state-space models to ZPK form.	Control System
zpkdata	Given a ZPK object, **zpkdata** extracts the zeros, poles, and gain and other information about the system.	Control System

Note: All the functions listed in the preceding table are included in *The Student Edition of MATLAB* with the following exceptions:

1. The RPI functions **bwcalc**, **cpole2t**, **rpole2t**, **small20**, **tstats**, and **vgain**.
2. The Control System Toolbox functions **ngrid** and **sgrid**.

TABLE B.1 *Where to find MATLAB commands described in the text and used in scripts*

Function	*Text*	*MATLAB Scripts*
*	17, 39, 44, 125	3.1, 3.3; 4.2; 6.4; 7.3; 8.1, 8.2; 9.1–9.5; 10.6
+	41, 44	3.2, 3.3; 4.2; 9.1
angle	86	
axis	84, 105, 146	5.1–5.3; 6.2, 6.3; 8.1–8.3
bode	100, 103	6.1, 6.2, 6.3; 7.3; 9.1–9.4; 10.6
bwcalc	125	7.3; 9.4
canon	73, 74	4.5
conv		3.2, 3.3, 3.5; 5.2, 5.3; 7.1–7.3; 8.1, 8.2; 9.5
cpole2t	15	2.4
ctrb		10.1
damp		6.3; 7.1
dcgain	44, 126	2.6; 3.3, 3.5; 6.3; 7.3; 8.2; 9.1–9.4; 10.6
deconv	129	
eig	72	4.5; 10.2, 10.3, 10.5, 10.6
feedback	46	3.4, 3.5; 4.2; 6.4; 7.1–7.3; 8.1, 8.2; –9.3, 9.4
find	21	2.7; 8.1; 9.1
findobj		4.3
get		2.1
hold	107	8.1, 8.2
impulse	13, 26	2.6, 2.8, 2.9; 4.4
interp1	159, 164	9.1–9.4
keyboard	5	
logspace	104	6.2, 6.3; 9.1, 9.3
lsim	70	2.7, 2.9; 4.4; 6.1; 7.2; 10.5, 10.6
margin	125, 138	6.4; 7.3; 9.1–9.4
ngrid	104	6.2
nichols	103	6.2
norm	185	10.3
nyquist	103, 104	6.2
obsv		10.4
place		10.2, 10.3, 10.5, 10.6
pole		2.8; 3.2–3.4; 4.3; 6.4; 10.6
print	5	
pzmap		2.1, 2.8, 2.9, 2.10; 4.3
rank		10.1, 10.4
reg	192	
residue	12	2.3–2.5; 7.1

Continues

APPENDIX B

Function	Text	MATLAB Scripts
reshape		6.3; 9.1, 9.3, 9.4
return	5	
rlocfind	82, 84, 135	5.1, 5.2; 8.1–8.3; 9.5
rlocus	82	5.1, 5.2; 8.1–8.3; 9.5
roots	26	2.8, 2.9; 3.2, 3.5; 5.3; 6.4; 9.5
rpole2t		2.4, 2.5
semilogx		6.3
set	83	5.1
ss	63, 64	4.1–4.5; 10.5, 10.6
sgrid		8.1
small20	207	
ssdata	62	
step		2.6, 2.10; 3.1–3.5; 4.4; 7.1; 8.1, 8.2; 9.1–9.5; 10.6
subplot	152, 197	
tf	8	2.1–2.3, 2.6–2.10; 3.1–3.5; 4.1; 5.2; 6.1–6.4; 7.1–7.3; 8.1, 8.2; 9.1–9.5
tfdata	8, 9	2.1, 2.9; 7.1
tstats	119, 135, 138	7.1; 8.2; 10.6
tzero		2.9; 4.3; 10.6
vgain	122	7.2
zpk	8, 64	2.2, 2.9; 3.1, 3.3; 4.1; 5.1, 5.2
zpkdata	8–10, 12, 44, 68	2.2, 2.3, 2.10; 3.1, 3.2, 3.4, 3.5; 4.3; 6.3

Annotated Bibliography

1. D'Azzo, John J., and Constantine Houpis, *Linear Control System Analysis and Design, Fourth Edition,* McGraw-Hill, New York, 1995.

2. Dorf, Richard C., and Robert Bishop, *Modern Control Systems, Eighth Edition,* Addison-Wesley, Reading, MA, 1998.

3. Franklin, Gene F., J. David Powell, and Abbas Emami-Naeini, *Feedback Control of Dynamic Systems, Third Edition,* Addison-Wesley, Reading, MA, 1994.

4. The MathWorks, *Control System Toolbox*, The MathWorks, Inc., Natick, MA, 1996.

5. Kuo, Benjamin C., *Automatic Control Systems, Seventh Edition,* Prentice-Hall, Upper Saddle River, NJ, 1995.

6. The MathWorks, *Using MATLAB version 5*, The MathWorks, Inc., Natick, MA, 1997.

7. Hanselman, Duane, and Bruce Littlefield, *The Student Edition of MATLAB, Version 5 User's Guide*, Prentice-Hall, Upper Saddle River, NJ, 1997.

8. Nise, Norman S., *Control Systems Engineering, Second Edition,* Benjamin/Cummings, Redwood City, CA, 1995.

9. Ogata, Katsuhiko, *Modern Control Engineering, Third Edition,* Prentice-Hall, Upper Saddle River, NJ, 1997.

10. Phillips, Charles L., and Royce D. Harbor, *Feedback Control Systems, Third Edition,* Prentice-Hall, Upper Saddle River, NJ, 1996.

11. Rohrs, Charles E., James L. Melsa, and Donald G. Schultz, *Linear Control Systems,* McGraw-Hill, New York, 1993.

12. Sigmon, Kermit, *MATLAB Primer, Fifth Edition,* CRC Press, Boca Raton, FL, 1998.

Index